Analysis and Control of Polynomial Dynamic Models with Biological Applications

Analysis and Control of Polynomial Dynamic Models with Biological Applications

Attila Magyar
University of Pannonia, Veszprém, Hungary

Gábor Szederkényi
Pázmány Péter Catholic University (PPKE), Budapest, Hungary

Katalin M. Hangos
University of Pannonia, Veszprém, Hungary
Computer and Automation Research Institute of Hungarian Academy of Sciences,
Budapest, Hungary

ACADEMIC PRESS

An imprint of Elsevier

Academic Press is an imprint of Elsevier
125 London Wall, London EC2Y 5AS, United Kingdom
525 B Street, Suite 1800, San Diego, CA 92101-4495, United States
50 Hampshire Street, 5th Floor, Cambridge, MA 02139, United States
The Boulevard, Langford Lane, Kidlington, Oxford OX5 1GB, United Kingdom

Library of Congress Cataloging-in-Publication Data
A catalog record for this book is available from the Library of Congress

British Library Cataloguing-in-Publication Data
A catalogue record for this book is available from the British Library

ISBN: 978-0-12-815495-3

For information on all Academic Press publications visit our
website at https://www.elsevier.com/books-and-journals

Working together
to grow libraries in
developing countries

www.elsevier.com • www.bookaid.org

Publisher: Candice Janco
Acquisition Editor: Glyn Jones
Editorial Project Manager: Naomi Robertson
Production Project Manager: Swapna Srinivasan

Typeset by SPi Global, India

For since the creation of the world God's invisible qualities—his eternal power and divine nature—have been clearly seen, being understood from what has been made, so that people are without excuse.

Romans 1:20

CONTENTS

Attila Magyar is an associate professor in the Department of Electrical Engineering and Information Systems at University of Pannonia. His background is information technology with an emphasis on systems and control theory. His research topics are nonlinear control and identification, and their application to process systems, smart electrical networks, and energetic systems.

Gábor Szederkényi is a professor in the Faculty of Information Technology and Bionics at the Pázmány Péter Catholic University, and a part-time scientific advisor at the Systems and Control Laboratory of the Institute for Computer Science and Control of the Hungarian Academy of Sciences. His background is computer engineering and nonlinear control. His main research interest is the computation-based analysis and control of nonlinear dynamical systems with special emphasis on models with biological and biochemical motivation.

Katalin Maria Hangos is a research professor at the Systems and Control Laboratory of the Research Institute for Computer Science and Control of the Hungarian Academy of Sciences. She is also a full professor in the Department of Electrical Engineering and Information Systems at University of Pannonia. She has a strong interdisciplinary background in systems and control theory and computer science, as well. Her main research interest lies in dynamic modeling of process systems for control and diagnostic purposes.

Quasipolynomial and kinetic systems constitute two important classes of nonnegative dynamic systems. Disciplines as diverse as chemistry and ecology make use of such systems to model nonlinear phenomena resulting from chemical/biochemical or biological interactions. In fact, as discussed in the introduction, such classes of polynomial dynamic systems can be considered universal descriptors of smooth nonlinear systems.

This book is the result of more than 10 years of the authors' research work in the field of quasipolynomial and kinetic systems; in combination with the authors' extensive experience in systems theory, this makes this book a valuable monograph on polynomial dynamic models. It provides a neat description of their underlying algebraic structure, as well as the structure of the associated graph representing systems' interactions. With that in hand, a comprehensive yet rigorous discussion of questions such as stability or dynamic equivalence under state-space or graph transformations is undertaken. Both concepts turn out to be fundamental in developing tools to guide structural or feedback control design so to make the system capable of performing prespecified dynamics, potentially linked to a particular functionality.

Of particular interest in the context of systems analysis and control is the extensive use of optimization to produce equivalent graph realizations of the system or to construct stabilizing feedback controllers. In this context, it is demonstrated in an instructive way that such problems can be systematically cast within the framework of linear programming or mixed-integer linear programming algorithms.

Not surprisingly, the topics covered are rather technical. However, the authors manage to make most of the material accessible by incorporating a good deal of well-placed examples, apart from a specific chapter that discusses a few relevant case studies in detail.

The book is to a large extent self-contained and only assumes basic knowledge of calculus, and some familiarity with standard notions from linear algebra. As written, it is oriented toward a broad community of graduate students and researchers in science and engineering, interested in exploring nonlinear phenomena from a systems theory perspective. Additionally, the present monograph will be particularly encouraging for chemical/biochemical engineers or applied mathematicians developing research in biological phenomena as well as theoretical biologists with interest in mathematical methods.

Prof. Antonio Alvarez Alonso

BioProcess Engineering Group, IIM-CSIC Vigo, Spain

September 2017

ACKNOWLEDGMENTS

We thank our coauthors and our students for the joint work and for the inspiring discussions. We are particularly indebted to Bernadett Ács, Attila Gábor, György Lipták, János Rudan, Zoltán Tuza, and Gergely Szlobodnyik (whose PhD project contributed the most to the content of this book).

The authors acknowledge the support of the National Research, Development, and Innovation Office—NKFIH through grants no. NF104706 and no. K115694.

A. Magyar was supported by the János Bolyai Research Scholarship of the Hungarian Academy of Sciences.

A. Magyar, G. Szederkényi, and K. M. Hangos

March 2017

CHAPTER 1

Introduction

The application of dynamical models describing the change of measured or computed quantities in time and/or in space has been indispensable not only in science but also in everyday life. It is also commonly accepted that dynamics plays a key role in understanding and influencing numerous complex processes taking place in living systems [1].

The deep understanding and the targeted manipulation of dynamical models' behavior are in the focus of systems and control theory that now provides us with really powerful methods for model analysis and controller synthesis in numerous engineering application fields. The efficient treatment of nonlinear and uncertain models is a well-developed field of control theory that has recently been a promising foundation for biological applications. Both system classes studied in the book, namely quasipolynomial (QP, or generalized Lotka-Volterra) systems and kinetic systems (also called [bio]chemical reaction networks, or simply CRNs) are so-called universal descriptors for smooth nonlinear systems. This means that they can represent all important dynamical phenomena that are present in biological (and also in general) dynamical systems. Moreover, both system classes are really natural descriptors of biological and biochemical processes: QP systems are the generalizations of Lotka-Volterra models originally used for modeling general population dynamics and related phenomena, while kinetic systems

Analysis and Control of Polynomial Dynamic Models with Biological Applications.
https://doi.org/10.1016/B978-0-12-815495-3.00001-8

composed of elementary reaction steps come from the description of (bio)chemical processes. The main practical advantage of QP and kinetic systems is their relatively simple matrix-based algebraic structure that allows the development of efficient computational (e.g., optimization-based) methods for their dynamical analysis and control. Moreover, the direct physical interpretation of many important system properties is often possible for these models.

Therefore, our aims with writing the book are the following: (1) To describe and illustrate the relation between dynamical, algebraic, and structural features of the studied model classes in a unique way not known by the authors in the literature. (2) To show the applicability of kinetic and QP representation in biological modeling and control through examples and case studies. (3) To show and emphasize the importance of quantitative models in understanding and influencing natural phenomena.

The target audience include graduate students in computer science, electrical, chemical, or bioengineering, applied mathematicians, and engineers, as well as researchers who are interested in a brief summary of the analysis and control of QP and kinetic systems.

1.1 THE NOTION AND SIGNIFICANCE OF DYNAMICAL MODELS IN THE DESCRIPTION OF NATURAL AND BIOLOGICAL PHENOMENA

Although the descriptive power of mathematical models is necessarily limited, their widespread utilization not only in research and development but also in the everyday life of today's technical civilization is clearly indispensable. When we are interested in the evolution of certain quantities, usually in time and/or space, we use dynamical models. The intensive use of dynamical models originates from classical physics for the description of the motion of objects. During the last century, by generalizing the concept of motion, the use of dynamical models became essential in other applied fields like electrical, chemical, and process engineering as well.

The key role of dynamics in the explanation of important phenomena occurring in living systems is now also a commonly accepted view. However, the possible complexity of processes and their interactions in living systems and the difficulties in measuring important quantities in vivo may still be an obstacle for the application of quantitative models of moderate size in life sciences. This situation has been continuously improving, since the accumulation of well-structured biological knowledge about molecular and higher level processes mainly in the form of reliable models, and the recent fast development in computer and computing sciences have converged, resulting in the birth of a new discipline called systems biology, that is now able to address important challenges in the field of life sciences. Dynamical models, for example, in the form of differential equations, are also routinely needed in synthetic biology, which is focused on the engineering design and construction of biological devices.

As indicated earlier, the theory of dynamical systems is an important link connecting different scientific fields with common tools, techniques, and a unified point of view in the convergence paradigm between life sciences, physics, and engineering [2]. This fact is clearly supported by numerous recent groundbreaking results heavily involving advanced computational models in biology, such as efficient defibrillation with a low shock intensity [3], the artificial pancreas operating under a wide range of conditions [4], or drug dosage control in cancer therapies [5], just to mention a few.

Within engineering, the area of systems and control theory deals with dynamic models for systems analysis and control, where the mathematical form of these models is a set of ordinary differential equations (abbreviated as ODEs) in the simplest cases. Nonlinear ODEs have a very good descriptive power, they can describe all the important dynamic phenomena that occur in a wide class of biological systems. Therefore, we restrict ourselves to *dynamic models in the form of ODEs in this book*. A brief summary of the notions and tools of systems and control theory used in this book is found in Appendix B.4.

It is well known that the methods and tools available for dynamic analysis and control of general nonlinear systems are computationally hard, if they exist at all. Therefore, one should use all the special properties of the system to be modeled in order to have efficient results. For biological applications, we most often have variables and parameters that are always nonnegative and differentiable, such as concentrations, temperatures, etc. Therefore, we shall focus on *smooth positive nonlinear ODE models* in this book.

1.2 KINETIC SYSTEMS AS GENERALIZATIONS OF REACTION KINETIC NETWORKS WITH MASS ACTION LAW

The dynamic description of kinetic systems originates from chemical reaction kinetics in physical chemistry. The first dynamic models of chemically reacting systems appear in the second half of the 19th century, when the factors affecting the rate of chemical reactions were discovered.

1.2.1 Chemical Reaction Networks With Mass Action Law

Gas phase is the simplest case in which reactions occur in a closed system under isothermal and isobaric conditions. Then the reaction rates are proportional to the concentration of the reactant chemical components, also called species, because reactions occur when the reactant molecules collide. This type of reaction kinetics is called *mass action law*.

A following simple example is to illustrate mass action law.

Example 1 (A simple reversible chemical reaction). Consider the simple chemical reaction

$$2H_2 + O_2 \rightarrow 2H_2O$$

and denote the concentration of hydrogen by x_1, that of oxygen by x_2, and that of the water by x_3. Assuming the presence of an inert gas in large extent, the dynamic equations that describe the time-evolution of the concentrations are as follows

$$\frac{dx_1}{dt} = -2kx_1^2 x_2$$

$$\frac{dx_2}{dt} = -kx_1^2 x_2$$

$$\frac{dx_3}{dt} = 2kx_1^2 x_2$$

where k is a positive parameter, the so-called reaction rate coefficient.

The example indicates that the dynamics are given in the mass action law case by a set of ODEs where the right-hand sides are polynomials. In addition, the reaction rates are negative for reactions that consume their reactants, resulting in the positivity of such systems.

Positive systems are characterized by the property that their differential variables remain positive during the dynamic evolution if one starts them from a positive initial condition. A more detailed formal introduction of positive systems is given in Section 2.1.2.

Plausible kinetic systems include physically meaningful chemical reactions which also obey mass conservation, that is, the molar weight of the species in both sides of the reaction is the same. Furthermore, the number of atoms of every kind is also the same in both sides of a reaction. For example, a chemical reaction

$$2\mathbf{X} \rightarrow \mathbf{X}$$

is not a physically meaningful chemical reaction, but its reaction rate may obey the mass action law.

Chemical reaction networks (abbreviated as CRNs) emerge from the theory of chemical kinetics with mass action law, assuming only the polynomial right-hand side and the positivity of the describing ODEs. The formal introduction of kinetic systems with mass action law will be given in Section 2.3.2.

1.2.2 Chemical Reaction Networks With Rational Functions as Reaction Rates

In practically important cases, however, the assumptions leading to mass action law kinetics rarely hold. Therefore, approximating reaction rate functions are used for describing the dynamics of pseudoreactions; these functions express the dynamic evolution of chemical processes consisting of a large number of elementary reaction steps under simplifying assumptions. The mathematical form of these reaction rate functions depends on the application area. In biochemical reaction networks, various combinations of Michaelis-Menten-type rate equations are most often used: these are rational functions where there is a reaction monomial in the nominator, and a positive polynomial in the denominator. This class of kinetic systems can be called *biochemical reaction networks*.

The following simple example illustrates the use of a rational function as a reaction rate expression.

Example 2 (A simple bio-reactor model). Consider a simple perfectly stirred reactor vessel under isothermal conditions, where there is a single feed-containing substrate solution. In the beginning a solution of biomass is put in the reactor, and we aim at producing biomass using fermentation.

The following notations are introduced

X : biomass concentration
S : substrate concentration
V : volume

F : inlet flow rate

S_F : inlet substrate concentration

$Y, \mu_{\max}, K_1, K_2$: kinetic parameters

The dynamics of the previous reactor can be described by the following set of ODEs

$$\frac{dX}{dt} = \mu(S)X - \frac{XF}{V}$$

$$\frac{dS}{dt} = -\frac{\mu(S)X}{Y} + \frac{(S_F - S)F}{V}$$

$$\text{where} \quad \mu(S) = \mu_{\max}\frac{S}{K_2 S^2 + S + K_1}$$

As we shall see later in Section 2.3.4, biochemical reaction networks are also positive systems, and their structure can also be described using the extended concepts and tools of chemical reaction networks with mass action law.

1.3 QP MODELS: UNIVERSAL DESCRIPTORS OF SMOOTH NONLINEAR SYSTEMS GENERALIZED FROM LOTKA-VOLTERRA SYSTEMS

Quasipolynomials are generalizations of ordinary polynomials which are allowed to have any real number in the exponents of the monomial terms. This way, a rich family of functions is created that has proven to be useful to represent the right-hand side of the dynamic ODE model equations of smooth nonlinear systems [6].

1.3.1 Original Lotka-Volterra Equations

The Lotka-Volterra equations, also known as predator-prey equations, are a pair of first-order, nonlinear, differential equations frequently used to describe the dynamics of biological systems in which two species interact, one as a predator and the other as prey. It is interesting to note that these equations appeared as the first qualitative dynamic models in biological ecology and were used to describe fish catches in the Adriatic Sea, and that the percentage of predatory fish caught had increased during the years of World War I (1914–18) [7].

Example 3 (Predator-prey system). Consider a closed ecological area in which predators and their prey coexist. Let us introduce the following notations

x : number of prey animals
y : number of predator animals
k : natural growth rate of preys when predators are missing
a : "meeting" rate of predators and preys
ℓ : natural death rate of predators when preys are missing
b : reproduction rate of predators when they eat a prey

The time evolution of the number of predators and preys are described by the simple bilinear set of ODEs as follows

$$\frac{dx}{dt} = k \cdot x - a \cdot x \cdot y$$
$$\frac{dy}{dt} = -\ell \cdot y + b \cdot x \cdot y \qquad (1.1)$$

It is easy to see that the Lotka-Volterra equations describe a positive system, and the time evolution of the variables x and y can be quite complex, showing oscillatory behavior depending on the values of the parameters.

1.3.2 Generalized Lotka-Volterra Equations

In the last decade of the 20th century, an increased interest has arisen to find general representations of smooth nonlinear dynamic models that are, on the one hand, flexible and general enough to have a good descriptive power for a wide class of nonlinearities and complex dynamical behavior, but on the other hand possess a treatable and well characterizable structure that can be used for model analysis. The class of QP systems have been proposed for this purpose [6] that is a set of ODEs in the following general form:

$$\frac{dx_i}{dt} = x_i \left(\lambda_i + \sum_{j=1}^{m} A_{i,j} \prod_{k=1}^{n} x_k^{B_{j,k}} \right), \quad i = 1, \ldots, n \qquad (1.2)$$

where both the parameters contained in the vector λ and matrix A, and the exponents represented as elements in the matrix B are real numbers.

Clearly, the Lotka-Volterra model in Eq. (1.1) is a special case of Eq. (1.2) with $n = m = 2$ and with $B = I$ where I is the unit matrix. Therefore, QP models are also called generalized LV models.

It turns out, the majority of smooth non-QP nonlinearity can be embedded into QP form [8], so QP models have a very good descriptive power. In addition, it was shown that the class of QP models can be split into equivalence classes using the so-called quasimonomial transformation (see later in Section 3.3 for more details), where an LV model serves as the representative of a class.

It is important to remark that kinetic systems both with mass action law and with rational function kinetics can be written in a QP form, while any LV model has a realization as a kinetic system with mass action law. The relationships between the kinetic and QP model classes will be investigated in details in Section 3.4.

A detailed formal introduction of the QP model class is given in Section 2.2.

CHAPTER 2

Basic Notions

2.1 GENERAL NONLINEAR SYSTEM REPRESENTATION IN THE FORM OF ODES

We will represent *autonomous dynamical models* in the form

$$\dot{x}(t) = f(x(t)), \quad x(0) = x_0 \tag{2.1}$$

where the independent variable $t \geq 0$ denotes time, $x(t) \in \mathbb{R}^n$ is the *state vector*, $f: \mathbb{R}^n \to \mathbb{R}^n$ is a smooth vector function, and $x_0 \in \mathbb{R}^n$ is the vector of

Analysis and Control of Polynomial Dynamic Models with Biological Applications.
https://doi.org/10.1016/B978-0-12-815495-3.00002-X

initial conditions. The time argument of x will be suppressed as it is common in the literature, that is, the ODE in Eq. (2.1) is simply written as

$$\dot{x} = f(x) \tag{2.2}$$

The simplest but practically still important case is when the *function f is smooth*, that is, it is (infinitely) many times continuously differentiable. Then, the solution of Eq. (2.1) for a given initial condition x_0 always exists (not necessarily for all $t \geq 0$), and it is unique. Moreover, the solutions depend continuously on the initial conditions [9].

A vector $x^* \in \mathbb{R}^n$ is called an *equilibrium point* of the system (2.1), if $f(x^*) = 0$. Clearly, the solutions of Eq. (2.1) remain constant at an equilibrium point, since the derivatives of all state variables are zero. The structure and properties of the (possibly multiple) equilibria of a general nonlinear system can be quite complex.

2.1.1 Autonomous Polynomial and Quasipolynomial Systems

When the function f is *linear*, the model (2.1) simplifies to a linear time invariant (LTI) autonomous system given by

$$\dot{x} = A \cdot x, \quad x(0) = x_0 \tag{2.3}$$

where $A \in \mathbb{R}^{n \times n}$. The dynamical properties and solutions of LTI systems are very well-known compared to the general nonlinear case [10].

It is immediately visible that in the case of LTI systems, all equilibria are characterized by the kernel of A.

2.1.1.1 Polynomial Systems

Let us introduce the notion of *multivariate monomial* $\varphi(x, m)$ for a state vector $x \in \mathbb{R}^n$ and exponent vector $m \in \overline{\mathbb{N}}_+^n$ in the form

$$\varphi(x, m) = x^m = \prod_{i=1}^{n} x_i^{m_i} \tag{2.4}$$

Then a polynomial system with multivariate polynomials $\mathcal{P}_j(x; k)$ in its right-hand sides is the ODE

$$\frac{dx_j}{dt} = \mathcal{P}_j(x; k) = \sum_{i=1}^{\ell} k_i \varphi_i(x, m^{(i)}), \quad j = 1, \ldots, n \tag{2.5}$$

2.1.1.2 Quasipolynomial Systems

One can generalize the notion of polynomial systems to allow nonintegers (i.e., rational numbers) in the exponents. Let $x, y \in \overline{\mathbb{R}}^n_+$. We will use the following notation

$$x^y = \prod_{i=1}^n x_i^{y_i} \tag{2.6}$$

where x^y is a quasimonomial. Then, *real multivariate quasipolynomial (QP) functions* are defined in the form

$$P(x) = \sum_{i=1}^m a_i x^{B^{(i)}} \tag{2.7}$$

where $a_i \in \mathbb{R}$ and $B^{(i)} \in \mathbb{R}^n$ is the ith column of matrix B for $i = 1, \ldots, m$.

If $B^{(i)} \in \overline{\mathbb{N}}^n_+$ in Eq. (2.7) $\forall i$, then P is called a *real multivariate polynomial function* or simply *polynomial function*. We recall that a function is called *smooth* if it has derivatives of all orders on its domain.

In order to have well-defined *quasimonomials* $x^{B^{(i)}}$ in Eq. (2.7), we usually restrict the domain of QP functions to the positive orthant (i.e., $x \in \mathbb{R}^n_+$). Naturally, polynomial functions are smooth, but QPs are generally not. In the positive orthant, however, QPs are also smooth functions.

The ODE

$$\frac{dx_j}{dt} = P_j(x) = \sum_{i=1}^m a_i x^{B^{(i)}}, \quad j = 1, \ldots, n \tag{2.8}$$

with QP functions $P_j(x)$ in its right-hand sides is an autonomous QP system.

2.1.1.3 Extension With Input Terms

We shall use the *input-affine extension of the autonomous dynamical models* of the dynamical model (2.1), that is in the following form

$$\dot{x}(t) = f(x(t)) + g(x(t))u(t), \quad x(0) = x_0 \tag{2.9}$$

where $u(t)$ is the *input vector*, and $g: \mathbb{R}^n \to \mathbb{R}^n$ is a smooth vector function. In order to have a positive polynomial or QP system, we require that the function $g(x)$ is also a polynomial or QP function, respectively.

2.1.2 Positive Polynomial Systems

A special and biologically important case emerges, when the dynamics (2.1) is invariant with respect to the nonnegative orthant $\overline{\mathbb{R}}_+^n$. This means that the coordinates of x remain nonnegative, if the entries of the initial condition vector are nonnegative. In such a case, the model (2.1) is called a *nonnegative system*. The necessary and sufficient conditions for establishing the nonnegative property are simple and will be given here based on [11]. The function f in Eq. (2.1) is called *essentially nonnegative* if whenever $x_i = 0$, then $f_i(x) \geq 0$ for all $x \in [0, \infty)^n$ for $i = 1, \ldots, n$. It is not difficult to prove that the system (2.1) is nonnegative if and only if f is essentially nonnegative.

A notable example of autonomous nonnegative systems is the class of Lotka-Volterra (LV) models defined as

$$\dot{x} = \text{diag}\{x\}(\Lambda + \mathcal{M}x), \quad x(0) = x_0 \qquad (2.10)$$

where $\Lambda \in \mathbb{R}^n$ and $\mathcal{M} \in \mathbb{R}^{n \times n}$. The individual differential equations corresponding to Eq. (2.10) are the following

$$\dot{x}_i = x_i \left(\Lambda_i + \sum_{j=1}^{n} \mathcal{M}_{ij} x_j \right), \quad i = 1, \ldots, n \qquad (2.11)$$

It is easy to see that the conditions of essential nonnegativity are fulfilled for the right-hand side of Eq. (2.11), since $x_i = 0$ implies $f_i(x) = 0$ for all i. More about this system class will be given in Section 2.2.

In the case of *polynomial systems*, however, the *conditions of nonnegativity* are more complicated. From the ODE form of polynomial systems in Eq. (2.5), the right-hand side functions $f_i(x)$ specialize to

$$f_i(x) = \sum_{j=1}^{\ell} k_j \prod_{\ell=1}^{n} x_\ell^{m_\ell^{(j)}}.$$

By setting $x_i = 0$, *one should have* $k_j \geq 0$ *for* $m_i^{(j)} = 0$ *for the nonnegativity condition above to be fulfilled.*

2.2 FORMAL INTRODUCTION OF THE QP MODEL FORM

Quasipolynomial systems form a notably wide class of nonnegative systems. This is explained by the fact that under mild conditions, the majority of smooth nonlinear systems can be embedded into a QP model form [12] (see details later in Section 3.4).

As we have already seen in Section 2.1.1, QPs are well-defined and are smooth functions, if one *restricts the domain of the state variables to the nonnegative orthant* $\overline{\mathbb{R}}_+^n$, that will be used here from now on.

2.2.1 QP Model Form

Let us have the vector of state variables x defined on the nonnegative orthant $\overline{\mathbb{R}}_+^n$. Then the QP models can be written in the form of the following ODE

$$\frac{dx_i}{dt} = x_i \left(\lambda_i + \sum_{j=1}^{m} A_{i,j} \prod_{k=1}^{n} x_k^{B_{j,k}} \right), \quad i = 1,\dots,n \qquad (2.12)$$

where $p_j = \prod_{k=1}^{n} x_k^{B_{j,k}}$ for $j = 1,\dots,m$ are the quasimonomials. Usually we assume $m \geq n$, that is, the number of quasimonomials is greater than that of the state variables. If this is not the case, then an equivalent model form obeying $m \geq n$ can be obtained by using model transformations [8] (see details on model transformations in Chapter 3).

The parameters of the QP model are the coefficient matrix $A \in \mathbb{R}^{n \times m}$, the exponent matrix $B \in \mathbb{R}^{m \times n}$, and the coefficient vector $\lambda \in \mathbb{R}^n$. It is commonly assumed that the exponent matrix B is of full rank (i.e., rank$(B) = n$). If this is not the case, then model transformations can bring the model into such an equivalent form [8].

It is easy to see that *the conditions of essential nonnegativity are fulfilled* for the right-hand side of Eq. (2.12), since $x_i = 0$ implies $f_i(x) = 0$ for all i, therefore QP systems are nonnegative systems.

The notion of autonomous QP systems has already been defined in Section 2.1.1 as an ODE with QP functions $P_j(x) = \sum_{i=1}^{m} a_i x^{B^{(i)}}$ in its right-hand sides. This general form can be easily transformed into the standard QP model form in Eq. (2.12), as it is illustrated in Example 4.

Example 4. Consider a simple polynomial, and thus QP model

$$\frac{dx_1}{dt} = 3x_1^2 + 2x_2 + 5, \quad \frac{dx_2}{dt} = 2x_2^2 + 3x_1 + 7x_2$$

Its standard QP model form, Eq. (2.12) will be

$$\frac{dx_1}{dt} = x_1(3x_1 + 2x_2 x_1^{-1} + 5x_1^{-1}), \quad \frac{dx_2}{dt} = x_2(2x_2 + 3x_1 x_2^{-1} + 7)$$

with the parameters

$$
\lambda = \begin{bmatrix} 0 \\ 7 \end{bmatrix}, \quad A = \begin{bmatrix} 3 & 2 & 5 & 0 & 0 \\ 0 & 0 & 0 & 2 & 3 \end{bmatrix}, \quad B = \begin{bmatrix} 1 & 0 \\ -1 & 1 \\ -1 & 0 \\ 0 & 1 \\ 1 & -1 \end{bmatrix}
$$

2.2.1.1 Compact Matrix-Vector Forms of QP Models
Let us introduce the notation diag$\{x\}$ for a diagonal matrix that has the elements x_i of the real vector $x \in \mathbb{R}^n$ in its diagonals. Then the matrix-vector form of the standard QP model (2.12) becomes

$$
\frac{dx}{dt} = \text{diag}\{x\}(\lambda + \mathcal{A}p), \quad x(0) = x_0 \tag{2.13}
$$

with $p_j = \prod_{k=1}^{n} x_k^{B_{j,k}}$, $j = 1, \ldots, m$ being the quasimonomials.

In some cases it is convenient to join the coefficients λ and $\mathcal{A}$ of a QP model. For this purpose we introduce the constant quasimonomial $p_0 = \underline{1} \in \mathbb{R}^m$, where $\underline{1}$ is a vector with all entries equal to 1. In addition let us introduce the element-wise logarithm operator Ln for an element-wise positive vector x as $[\text{Ln}(x)]_i = \ln x_i$. Then, the relationship between the state variables and the quasimonomials can be written in the following compact form:

$$
\text{Ln}(p) = \mathcal{B}\,\text{Ln}(x) \tag{2.14}
$$

Furthermore, we extend the original exponent matrix $\mathcal{B}$ with a new zero first row to obtain $\hat{\mathcal{B}}$, and the original coefficient matrix $\mathcal{A}$ with a new first column being equal to λ to obtain $\hat{\mathcal{A}}$. Similarly, the extended quasimonomial vector $\hat{p} \in \mathbb{R}^{m+1}$ will have a new first element being equal to 1, and the rest shifted. Then the compact matrix-vector form of a QP model is as follows

$$
\frac{dx}{dt} = \text{diag}\{x\}\hat{\mathcal{A}} \cdot \hat{p}, \quad x(0) = x_0 \tag{2.15}
$$

2.2.1.2 An Entropy-Like Lyapunov Function Candidate for QP Models
Similar to the case of linear time-invariant systems, where a quadratic Lyapunov function candidate exists, there is a well-known candidate Lyapunov function family for QP systems [13, 14], which is in the form:

$$V(p) = \sum_{i=1}^{m} c_i \left(p_i - p_i^* - p_i^* \ln \frac{p_i}{p_i^*} \right) \tag{2.16}$$

where p is the vector of quasimonomials.

2.2.2 LV Systems

A remarkable and important special case of QP models is when the exponent matrix $\mathcal{B} \in \mathbb{R}^{m \times n}$ is square (i.e., $m = n$), and it is the unit matrix (i.e., $\mathcal{B} = I$). In this case the coefficient matrix $\mathcal{A}$ is also square: this is called a Lotka-Volterra (abbreviated as LV) model

$$\frac{dx}{dt} = \operatorname{diag}\{x\}(\Lambda + \mathcal{M}x), \quad x(0) = x_0 \tag{2.17}$$

where $\Lambda \in \mathbb{R}^m$ and $\mathcal{M} \in \mathbb{R}^{m \times m}$ is the LV coefficient matrix. It is important to remark that the state variables are identical to the quasimonomials in this special case.

As we shall see later in Section 3.3.2, any QP model can be transformed to an LV form by embedding and nonlinear state transformation, but in this case the coefficient matrix $\mathcal{M}$ will not be in general of full rank.

An excellent summary on the qualitative dynamical properties of LV systems can be found in [15], and this is the subject of Section 4.1.2.

2.2.3 Extension With Input Term

The simplest and yet widely applied way of extending the original QP model (2.13) with an input-affine term is to consider a *linear input structure*, that can be formally derived by regarding λ as a function of the input vector $u \in \mathbb{R}^\mu$ such that the input-affine state equation is in the form

$$\frac{dx}{dt} = \operatorname{diag}(x)\,(\mathcal{A}p + \phi\,u) \tag{2.18}$$

where $p_j = \prod_{k=1}^{n} x_k^{\mathcal{B}_{j,k}}$ for $j = 1, \ldots, m$ are the quasimonomials, and $\phi \in \mathbb{R}^{n \times \mu}$ is a constant matrix.

This structure will be used later in Section 5.1 to design stabilizing feedback controllers to QP and LV systems.

2.3 INTRODUCTION OF KINETIC MODELS WITH MASS ACTION AND RATIONAL REACTION RATES

It is important to remark here that similar to [16, 17], or [18], we introduce and treat reaction networks as a general system class describing nonlinear dynamical systems with nonnegative states. Therefore, generally we do not require that they fulfill actual physicochemical constraints like mass conservation. Models that are physically realizable and satisfy fundamental physicochemical properties are naturally special cases of the introduced kinetic models.

2.3.1 General Notions for Reaction Networks

The basic building blocks of CRN models are *elementary reaction steps* of the form

$$C_i \to C_j \tag{2.19}$$

denoting chemical complex C_i transforming into complex C_j in the network.

Using the previous notion of elementary reactions, we can give the formal definition of CRN-related notions. The triplet $(\mathcal{S}, \mathcal{C}, \mathcal{R})$ is called a *chemical reaction network structure* or simply *CRN structure*, where the components are defined as

- A set of *species*: $\mathcal{S} = \{X_i \mid i \in \{1, \ldots, n\}\}$
- A set of *complexes*: $\mathcal{C} = \{C_j \mid j \in \{1, \ldots, m\}\}$, where

$$C_j = \sum_{i=1}^{n} \alpha_{ji} X_i \quad j \in \{1, \ldots, m\}$$

$$\alpha_{ji} \in \overline{\mathbb{N}}_+^n \quad j \in \{1, \ldots, m\}, \ i \in \{1, \ldots, n\}$$

The complexes are formal linear combinations of the species with nonnegative integer coefficients, called the *stoichiometric coefficients*.

- A set of *reactions*: $\mathcal{R} \subseteq \{(C_i, C_j) \mid C_i, C_j \in \mathcal{C}\}$.

  The ordered pair (C_i, C_j) corresponds to the reaction $C_i \to C_j$, where C_i is called the *source* (or *reactant*) *complex* and C_j is the *product complex*.

The *complex composition* matrix $Y \in \mathbb{R}^{n \times m}$ contains the stoichiometric coefficients as

$$Y_{ij} = \alpha_{ji}, \quad i = 1, \ldots, n, \ j = 1, \ldots, m \tag{2.20}$$

The concentration of species X_i will be denoted by x_i for $i = 1, \ldots, n$. The dynamics (i.e., the change of the concentrations in time) of a CRN can be described using the following set of ODEs

$$\dot{x} = \sum_{(C_i, C_j) \in \mathcal{R}} (Y_{\cdot j} - Y_{\cdot i}) \cdot R_{ij}(x) \tag{2.21}$$

where $R_{ij} \colon \overline{\mathbb{R}}_+^n \mapsto \overline{\mathbb{R}}_+$ is the *reaction rate function* corresponding to the reaction $C_i \to C_j$. Let r be the number of reactions in the CRN (i.e., $r = |\mathcal{R}|$). Then, using multiindices for the complexes, the reactions in the CRN can be listed as

$$C_{i_k} \to C_{j_k}, \quad \text{for} \quad k = 1, \ldots, r \tag{2.22}$$

with i_k and j_k being the indices of the reactant and product complexes of the kth reaction, respectively. We can easily construct a vector function $R \colon \overline{\mathbb{R}}_+^n \mapsto \overline{\mathbb{R}}_+^r$ from the individual reaction rate functions as

$$R(x) = [R_{i_1 j_1}(x) \; R_{i_2 j_2}(x) \; \ldots \; R_{i_r j_r}(x)]^T \tag{2.23}$$

where $R_{i_k j_k}$ is the reaction rate function corresponding to the kth reaction for $k = 1, \ldots, r$. Using these notations, Eq. (2.21) can be rewritten as

$$\dot{x} = \mathbf{S} \cdot R(x) \tag{2.24}$$

where $\mathbf{S} \in \mathbb{R}^{n \times r}$ is the *stoichiometric matrix* with $\mathbf{S}_{\cdot, k} = Y_{\cdot j_k} - Y_{\cdot i_k}$ for $k = 1, \ldots, r$. The columns of $\mathbf{S}$ are called the *reaction vectors*, and they span the so-called *stoichiometric subspace* S, that is,

$$S = \mathrm{im}(\mathbf{S}) \tag{2.25}$$

Integrating Eq. (2.24) gives

$$x(t) = x(0) + \sum_{j=1}^{r} \left(\int_0^t R_j(x(\tau)) d\tau \right) \mathbf{S}_{\cdot j} \tag{2.26}$$

where R_j denotes the jth coordinate function of R defined in Eq. (2.23). Therefore, the states always remain in the so-called *stoichiometric compatibility class* corresponding to the initial conditions and defined by

$$\mathcal{X}_{x(0)} = \{x(0) + v \mid v \in \mathrm{im}(\mathbf{S})\} \tag{2.27}$$

Note that the general definition of a reaction network structure in the form of $\mathcal{N} = (\mathcal{S}, \mathcal{C}, \mathcal{R})$ encodes only the complex composition and the network structure (i.e., the interactions), but not the reaction rates that are otherwise necessary to write the dynamical equations (2.21). In the following, we will introduce two cases of CRN representation based on the form of the reaction rate functions.

2.3.1.1 Reaction Graph
Based on the previous description, it is straightforward to assign a directed graph called the *reaction graph* (or *Feinberg-Horn-Jackson* graph) to a reaction network structure as follows.

- The *vertices* correspond to the complexes, $V(G) = C$.
- The *directed edges* describe the reactions, $E(G) = \mathcal{R}$. There is a directed edge from vertex C_i to vertex C_j if and only if the reaction $C_i \rightarrow C_j$ takes place in the reaction network.

Loops are not allowed in a reaction graph. The so-called *linkage classes* are the weak components of the reaction graph.

One can characterize the reaction graph using its incidence matrix $B_G \in \{-1, 0, 1\}^{m \times r}$ where r is the number of reactions. Each reaction in the CRN is represented by the appropriate column of B_G as follows. Let the ℓth reaction in the CRN be $C_j \rightarrow C_i$ for $1 \leq \ell \leq r$. Then the ℓth column vector of B_G is characterized as: $[B_G]_{i\ell} = 1$, $[B_G]_{j\ell} = -1$, and $[B_G]_{k\ell} = 0$ for $k = 1, \ldots, r, k \neq i, j$.

Using the incidence matrix, the kinetic dynamics (2.21) can be alternatively rewritten as

$$\dot{x} = Y \cdot B_G \cdot R(x) \tag{2.28}$$

that is, the stoichiometric matrix is given by $\mathbf{S} = Y \cdot B_G$.

2.3.1.2 Important Structural Properties of Reaction Networks
It is possible to utilize certain structural and/or parameter-dependent properties of reaction kinetic systems that support the dynamical analysis of the system. For this, we introduce a couple of notions.

A reaction graph is *weakly reversible* if each weak component in it is actually a strong component. This means that if there exists a directed from C_i to C_j, then there exists a directed path to C_j to C_i. Weak reversibility is also equivalent to the property that each directed edge is lying on a directed cycle in the graph.

A reaction graph is *(fully) reversible* if for each reaction $C_i \rightarrow C_j$, there exists the opposite reaction $C_j \rightarrow C_i$.

Deficiency denoted by δ is an important nonnegative integer number characterizing the stoichiometry and structure of a reaction network. There are several equivalent ways to define δ. We first introduce the following definition that can be found, for example, in [19]:

$$\delta = \dim(\ker(Y) \cap \mathrm{im}(B_G)) \tag{2.29}$$

The original definition of deficiency (used, e.g., in [16]) is

$$\delta = m - l - s \tag{2.30}$$

where m is the number of complexes, l is the number of linkage classes, and s is the dimension of the stoichiometric subspace.

2.3.2 Reaction Networks With Mass Action Kinetics

The physical origin of chemical reaction network (CRN) models obeying the mass action law is found in the molecular collision picture of gas phase chemical reactions, where the reaction rate is proportional to the number of collisions, that is, to the concentration of the reactants. Using this consideration, the reaction rate R_{ij} corresponding to the elementary reaction step (2.19) is given by

$$R_{ij}(x) = k_{ij} \prod_{k=1}^{n} x_k^{\alpha_{ik}} \tag{2.31}$$

where k_{ij} is the *reaction rate coefficient*. By convention, $k_{ij} = 0$ will comfortably denote that $(C_i, C_j) \notin \mathcal{R}$. Therefore, $k_{ij} > 0$ for all i,j for which $(C_i, C_j) \in \mathcal{R}$. It is easy to see that using $\mathcal{S}, \mathcal{C}, \mathcal{R}$, and the reaction rate coefficients completely characterize the stoichiometry, structure, and dynamics of the network. Therefore, the quadruple $\mathcal{N} = (\mathcal{S}, \mathcal{C}, \mathcal{R}, \mathcal{K})$ will be called a *mass action reaction network*, where $\mathcal{K}$ denotes the set of rate coefficients (i.e., $\mathcal{K} = \{k_{ij} \mid i,j = 1, \ldots, m\}$).

Due to Eqs. (2.20), (2.31), and (2.24), the dynamics of mass action systems can be characterized by two matrices. One of them is the complex composition matrix Y defined previously, while the other is the so-called *Kirchhoff matrix*. $A_k \in \mathbb{R}^{m \times m}$ is the *Kirchhoff matrix* of a CRN if its entries are determined by the reaction rate coefficients as follows:

$$[A_k]_{ij} = \begin{cases} k_{ji} & \text{if } i \neq j \\ -\sum_{l=1, l \neq i}^{m} k_{il} & \text{if } i = j \end{cases} \tag{2.32}$$

Since the sum of entries in each column is zero, A_k is a so-called *column conservation matrix*. Furthermore, it is immediately visible that A_k is a compartmental Metzler matrix with the properties described in Section B.2.

Assuming mass-action kinetics, the dynamics of the CRN can be described by dynamical equations of the following form

$$\dot{x} = Y \cdot A_k \cdot \psi(x) \tag{2.33}$$

where $\psi \colon \mathbb{R}^n_+ \to \mathbb{R}^m_+$ is a monomial-type vector-mapping,

$$\psi_j(x) = \prod_{i=1}^{n} x_i^{\alpha_{ji}}, \quad j = 1, \ldots, m \tag{2.34}$$

2.3.2.1 The Reaction Graph of Mass Action Networks
The weighted directed reaction graph $G(V, E)$ of mass action networks with the edge weighting $w \colon E(G) \to \mathbb{R}_+$ is defined as follows.

- The *vertices* correspond to the complexes, $V(G) = \mathcal{C}$.
- The *directed edges* describe the reactions, $E(G) = \mathcal{R}$.
     There is a directed edge from vertex C_i to vertex C_j if and only if the reaction $C_i \to C_j$ takes place (i.e., $k_{ij} > 0$).
- the *weights* of the edges are the reaction rate coefficients, $w(C_i, C_j) = k_{ij}$ where $(C_i, C_j) \in \mathcal{R}$.

Loops and multiple edges are not allowed in the reaction graph of mass action systems. It can be seen from the previous definition that the Kirchhoff matrix A_k is in fact the negative transpose of the weighted Laplacian matrix of the reaction graph.

Example 5 (Mass action model of the simple irreversible Michaelis-Menten kinetics). The so-called Michaelis-Menten kinetics is a fundamental type of biochemical reactions that occur frequently in applications. Reflecting the usual conditions of enzymatic kinetics, we consider a closed isothermal system with constant physicochemical properties.

Then, the overall reaction equation of Michaelis-Menten kinetics is

$$\mathbf{E} + \mathbf{S} \longrightarrow \mathbf{E} + \mathbf{P} \tag{2.35}$$

where **E** is the enzyme, **S** is the substrate, and **P** is the product of the reaction. The enzyme acts as a catalyzator. The mechanism assumes that the substrate forms a complex **ES** with the enzyme in a reversible reaction step that can react to form the product **P** and giving back the enzyme **E**.

The simplest mechanism with two reactions obeying the mass action law is

$$\mathbf{E + S \rightleftarrows ES \quad ES \rightleftarrows E + P} \tag{2.36}$$

The full reaction kinetic model consists of the component mass balances for all of the species **S**, **E**, **ES**, and **P**:

$$\frac{d[S]}{dt} = -k_1^+[E][S] + k_1^-[ES] \tag{2.37}$$

$$\frac{d[E]}{dt} = -k_1^+[E][S] + k_1^-[ES] - k_2^-[E][P] + k_2^+[ES] \tag{2.38}$$

$$\frac{d[ES]}{dt} = +k_1^+[E][S] - k_1^-[ES] + k_2^-[E][P] - k_2^+[ES] \tag{2.39}$$

$$\frac{d[P]}{dt} = -k_2^-[E][P] + k_2^+[ES] \tag{2.40}$$

where $[C]$ denotes the concentration of the specie **C**. Note that in most of the cases the second reaction step is assumed to be irreversible (i.e., $k_2^- = 0$ in the previous model). This case will be called *irreversible simple Michaelis-Menten mechanism*.

The previous simple mechanism can be depicted using the so called *reaction graph* seen in Fig. 2.1, that is a weighted directed graph with the complexes as nodes, and the reactions as directed edges. The reaction rate coefficients $k_i^{+,-}$ are the edge weights.

Let us introduce the standard notations to describe the irreversible mechanism as follows. Let

$$X_1 = [S], \quad X_2 = [E], \quad X_3 = [ES], \quad X_4 = [P]$$

Then, the complexes of the system are in the set $\mathcal{C} = \{C_1, C_2, C_3\}$, where

$$C_1 = X_1 + X_2, \quad C_2 = X_3, \quad C_3 = X_2 + X_4 \tag{2.41}$$

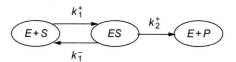

Fig. 2.1 *The reaction graph of the mass action-type model of the simple irreversible Michaelis-Menten kinetics.*

This defines the complex composition matrix Y and the corresponding monomial vector ψ as

$$Y = \begin{bmatrix} 1 & 0 & 0 \\ 1 & 0 & 1 \\ 0 & 1 & 0 \\ 0 & 0 & 1 \end{bmatrix}, \quad \psi(x) = \begin{bmatrix} x_1 x_2 \\ x_3 \\ x_2 x_4 \end{bmatrix}$$

Let us consider the reactions in the following order:

$$C_1 \rightarrow C_2, \quad C_2 \rightarrow C_1, \quad C_2 \rightarrow C_3$$

Thus the reaction rate vector is

$$R(x) = [k_1^+ x_1 x_2 \quad k_1^- x_3 \quad k_2^+ x_3]^T$$

and the stoichiometric matrix is given by

$$S = \begin{bmatrix} -1 & 1 & 0 \\ -1 & 1 & 1 \\ 1 & -1 & -1 \\ 0 & 0 & 1 \end{bmatrix}$$

Using Eq. (2.24), we obtain the ODEs of the system dynamics as

$$\begin{aligned} \dot{x}_1 &= -k_1^+ x_1 x_2 + k_1^- x_3 \\ \dot{x}_2 &= -k_1^+ x_1 x_2 + (k_1^- + k_2^+) x_3 \\ \dot{x}_3 &= k_1^+ x_1 x_2 - (k_1^- + k_2^+) x_3 \\ \dot{x}_4 &= k_2^+ x_3 \end{aligned} \tag{2.42}$$

The Kirchhoff matrix encoding the reaction graph is

$$A_k = \begin{bmatrix} -k_1^+ & k_1^- & 0 \\ k_1^+ & -(k_1^- + k_2^+) & 0 \\ 0 & k_2^+ & 0 \end{bmatrix}$$

while the incidence matrix is

$$B_G = \begin{bmatrix} -1 & 1 & 0 \\ 1 & -1 & -1 \\ 0 & 0 & 1 \end{bmatrix}$$

It is easy to check that all the introduced descriptions (2.24), (2.28), and (2.33); that is, $\dot{x} = Y \cdot A_k \cdot \psi(x)$, $\dot{x} = Y \cdot B_G \cdot R(x)$, and $\dot{x} = S \cdot R(x)$ give the ODE system (2.42).

2.3.2.2 Important Properties of Mass Action-Type Reaction Networks and Their Implications

The reversibility and deficiency structural properties introduced earlier in this chapter in Section 2.3.1.2 have important implication to the qualitative dynamical properties of mass action CRNs formulated in the celebrated *Deficiency Zero Theorem* published in [16].

To see this we first recall from the literature [20] that a mass action CRN is weakly reversible if and only if there exists a vector χ^* with strictly positive elements in the kernel of A_k, that is, there exists $b \in \mathbb{R}_+^n$ such that $A_k \cdot b = 0$. This means that *this positive vector χ^* gives rise to a positive equilibrium point of the dynamic model of the CRN.* With some abuse of the notions, we will call a Kirchhoff matrix A_k *weakly reversible* if the reaction graph corresponding to A_k is weakly reversible.

Then the statements of the well-known *Deficiency Zero Theorem* published in [16] can be summarized as follows:

(i) If a deficiency zero (not necessarily mass action) network is not weakly reversible, then it cannot have a strictly positive equilibrium point. Moreover, the state variables cannot follow a strictly positive cyclic trajectory in this case.

(ii) A deficiency zero weakly reversible reaction network with mass action kinetics has precisely one strictly positive equilibrium point in each so-called stoichiometric compatibility class that is at least locally stable with a known logarithmic Lyapunov function, irrespectively of the values of the reaction rate coefficients.

The Lyapunov function mentioned in point (ii) has the following form

$$V(x) = \sum_{i=1}^{n} x_i \left(\ln \left(\frac{x_i}{x_i^*} - 1 \right) \right) + x_i^* \qquad (2.43)$$

where x^* denotes the positive equilibrium point of the system.

More about the Deficiency Zero Theorem will follow in Section 4.2 when discussing the stability of kinetic systems.

The following two notions originally come from the thermodynamic analysis of chemical reaction network models. A mass action CRN is called *complex balanced* if there exists $x^* \in \mathbb{R}_+^n$ for which

$$A_k \cdot \psi(x^*) = 0 \qquad (2.44)$$

It is immediately visible that x^* is an equilibrium point if Eq. (2.44) is fulfilled. It is known from the literature that if Eq. (2.44) is satisfied for any $x^* \in \mathbb{R}_+^n$ then it is fulfilled for all other equilibria of the system. Therefore, complex balance is indeed the property of the CRN and not only that of an equilibrium point. It is also easy to see from Eq. (2.44) that the signed sum of the rates of incoming and outgoing reactions for each complex in the reaction graph is zero at any equilibrium point for complex balanced networks. Moreover, *complex balance implies weak reversibility*, since there exists a strictly positive vector in the kernel of A_k.

A reversible mass action CRN is called *detailed balanced* if there exists $x^* \in \mathbb{R}_+^n$ so that

$$R_{ij}(x^*) = R_{ji}(x^*) \quad \text{for} \quad i,j \text{ such that } (C_i, C_j) \in \mathcal{R} \qquad (2.45)$$

Eq. (2.45) shows that the forward and reverse reaction rates are pairwise equal at x^*. Therefore, *detailed balance implies complex balance*, and x^* is an equilibrium point of the dynamics. It is also true in this case that if Eq. (2.45) is fulfilled for any $x^* \in \mathbb{R}_+^n$, then it is satisfied for all other equilibria of the CRN.

It is visible from the previous definitions that *weak reversibility is a property of the reaction graph alone, while deficiency depends both on the graph structure and on the stoichiometry of the reaction network.* Detailed balance and complex balance depend additionally on the network parameters (rate coefficients). Interestingly, as the next section will show, all these important properties are generally not encoded into the kinetic differential equations, and different network structures/parametrizations may correspond to the same dynamics.

2.3.3 Kinetic Realizability and Structural Nonuniqueness of Mass Action-Type Reaction Networks

According to Eqs. (2.33), (2.34) the dynamics of a reaction network can be given by a set of polynomial ODEs, but not every polynomial system describes a CRN. Therefore, we will first shortly summarize the conditions for representing a polynomial dynamical system as a CRN.

Let $x: \mathbb{R} \to \mathbb{R}_+^n$ be a function, $M \in \mathbb{R}^{n \times p}$ a matrix, and $\varphi: \mathbb{R}_+^n \to \mathbb{R}_+^p$ a monomial function. The polynomial system

$$\dot{x} = M \cdot \varphi(x) \qquad (2.46)$$

is called *kinetic* if there exist a matrix $Y \in \overline{\mathbb{N}}_+^{n \times m}$ and a Kirchhoff matrix $A_k \in \mathbb{R}^{m \times m}$, so that

$$M \cdot \varphi(x) = Y \cdot A_k \cdot \psi(x) \qquad (2.47)$$

where $\psi \colon \mathbb{R}_+^n \to \mathbb{R}_+^m$ is a monomial function determined by the entries of matrix Y, $\psi_j(x) = \prod_{i=1}^n x_i^{[Y]_{ij}}$ for $j \in \{1, \ldots, m\}$. The problem of kinetic realizability of polynomial ODE models was first examined and solved in [21] where it was shown that the necessary and sufficient condition for kinetic realizability of a polynomial vector field is that all coordinates functions of f in Eq. (2.1) must have the form

$$f_i(x) = -x_i g_i(x) + h_i(x), \quad i = 1, \ldots, n \qquad (2.48)$$

where g_i and h_i are polynomials with nonnegative coefficients. It is easy to prove that kinetic systems are nonnegative [11].

2.3.3.1 Procedure for Computing a Canonical Mechanism
In [21], a realization procedure is presented for producing the so-called *canonical mechanism* from a set of kinetic ODEs. We will briefly outline this procedure in following. Let us write the polynomial coordinate functions of the right-hand side of a kinetic system (2.2) as

$$f_i(x) = \sum_{j=1}^{r_i} m_{ij} \prod_{k=1}^n x^{b_{jk}} \qquad (2.49)$$

where r_i is the number of monomial terms in f_i. Let us denote the transpose of the ith standard basis vector in $\mathbb{R}^n$ as e_i^T and let $B_j = [b_{j1} \ \ldots \ b_{jn}]$. Then the steps for producing the canonical mechanism realizing the dynamics (2.2) are the following.

Procedure 1 (For constructing the canonical mechanism [21]).

For each $i = 1, \ldots, n$ and for each $j = 1, \ldots, r_i$ do:

1. $C_j = B_j + \operatorname{sign}(m_{ij}) \cdot e_i^T$
2. Add the following reaction to the graph of the realization

$$\sum_{k=1}^n b_{jk} X_k \longrightarrow \sum_{k=1}^n c_{jk} X_k \qquad (2.50)$$

with reaction rate coefficient $|m_{ij}|$, where $C_j = [c_{j1} \ \ldots \ c_{jn}]$.

It is visible that in each step a new reaction is assigned to each monomial of Eq. (2.49).

2.3.3.2 Dynamic Equivalence

The matrices Y and A_k characterize the dynamics of the kinetic system, as well as the CRN. However, the polynomial kinetic system does not uniquely determine the matrices. Reaction networks with different sets of complexes and reactions can be governed by the same dynamics (see e.g., [22–24]). If the matrices Y and A_k of a reaction network fulfill Eq. (2.47), then the pair (Y, A_k) (and the corresponding CRN) is called a *dynamically equivalent realization* of the kinetic system (2.46).

Efficient numerical optimization-based methods exist [24] to determine dynamically equivalent realizations of a given kinetic system model that can be used in dynamical analysis and controller design for such systems. These methods will be described in details later in Section 4.5. The following simple example illustrates the notion of dynamically equivalent realizations.

Example 6. Let us consider the irreversible Michaelis-Menten system from Example 5 again. Using the notation in Eq. (2.46), the kinetic equations (2.42) can be given with the following matrix M and monomial function φ:

$$
M = \begin{bmatrix} -k_1^+ & k_1^- \\ -k_1^+ & k_1^- + k_2^+ \\ k_1^+ & -(k_1^- + k_2^+) \\ 0 & k_2^+ \end{bmatrix}, \quad \varphi(x) = \begin{bmatrix} x_1 x_2 \\ x_3 \end{bmatrix} \tag{2.51}
$$

Running Procedure 1 on Eqs. (2.42) gives the canonical reaction network that is depicted in Fig. 2.2. Note that the set of complexes in this case is different from the ones that appeared in Example 5, and the canonical realization contains only C_1 and C_2 from the original system listed in Eq. (2.41). Thus the example clearly shows that the "true" complex set generally cannot be retrieved from the dynamical description alone.

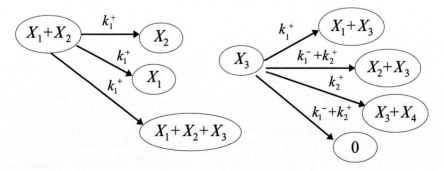

Fig. 2.2 The canonical realization of the ODE system corresponding to the simple irreversible Michaelis-Menten kinetics.

Let us number the complexes of the canonical realization as follows

$$C_1 = X_1 + X_2, \; C_2 = X_3, \; C_3 = X_2, \; C_4 = X_1, \; C_5 = X_1 + X_2 + X_3,$$
$$C_6 = X_1 + X_3, \; C_7 = X_2 + X_3, \; C_8 = 0, \; C_9 = X_3 + X_4$$

Then the matrices characterizing the network are

$$Y = \begin{bmatrix} 1 & 0 & 0 & 1 & 1 & 1 & 0 & 0 & 0 \\ 1 & 0 & 1 & 0 & 1 & 0 & 1 & 0 & 0 \\ 0 & 1 & 0 & 0 & 1 & 1 & 1 & 0 & 1 \\ 0 & 0 & 0 & 0 & 0 & 0 & 0 & 0 & 1 \end{bmatrix}$$

$$A_k = \begin{bmatrix} -3k_1^+ & 0 & 0 & 0 & 0 & 0 & 0 & 0 & 0 \\ 0 & -3k_1^- - 3k_2^+ & 0 & 0 & 0 & 0 & 0 & 0 & 0 \\ k_1^+ & 0 & 0 & 0 & 0 & 0 & 0 & 0 & 0 \\ k_1^+ & 0 & 0 & 0 & 0 & 0 & 0 & 0 & 0 \\ k_1^+ & 0 & 0 & 0 & 0 & 0 & 0 & 0 & 0 \\ 0 & k_1^- & 0 & 0 & 0 & 0 & 0 & 0 & 0 \\ 0 & k_1^- + k_2^+ & 0 & 0 & 0 & 0 & 0 & 0 & 0 \\ 0 & k_1^- + k_2^+ & 0 & 0 & 0 & 0 & 0 & 0 & 0 \\ 0 & k_2^+ & 0 & 0 & 0 & 0 & 0 & 0 & 0 \end{bmatrix}$$

and the monomial function is given by

$$\psi(x) = [x_1 x_2 \; x_3 \; x_2 \; x_1 \; x_1 x_2 x_3 \; x_1 x_3 \; x_2 \; x_3 \; 1 \; x_3 x_4]^T.$$

It can be checked that the product $Y \cdot A_k \cdot \psi(x) = M \cdot \varphi(x)$. Therefore, both reaction networks visible in Figs. 2.1 and 2.2 are dynamically equivalent realizations of the polynomial system $\dot{x} = M \cdot \varphi(x)$ defined by Eq. (2.51).

2.3.4 Reaction Networks With Rational Function Kinetics

In this section, we extend the previous reaction network class to models having rational functions in the reaction rates. The description is based on [25, 26].

In this case we allow that a complex may react to other complexes with several different (not necessarily mass action type) reaction rates. Therefore, the reaction rate functions have the form $R_{ijl}: \overline{\mathbb{R}}_+^n \mapsto \overline{\mathbb{R}}_+$, where i, j, and l correspond to the source complex, product complex, and kinetics, respectively. In general, it is possible that the species in a complex give rise to multiple reaction paths, thus multiple different reaction rate functions may be assigned to each complex. These reaction rates can be categorized based on their mathematical formulations. Using the notations of complexes and kinetics, the elementary reaction steps can be written as $C_i \xrightarrow{R_{ijl}} C_j$, where the reaction rate R_{ijl} of the reaction is decomposed as

$$R_{ijl}(x) = k_{ijl} \cdot g_{il}(x) \tag{2.52}$$

In the earlier equation, $k_{ijl} \in \mathbb{R}_+$ is a constant, nonnegative *principal reaction rate coefficient* and g_{il} is a function of the species concentrations. Similar to Eq. (2.23) the reaction rate vector can be written as

$$R(x) = [R_{i_1 j_1 l_1}(x) \ R_{i_2 j_2 l_2}(x) \ \dots \ R_{i_r j_r l_r}(x)]^T \tag{2.53}$$

where $R_{i_k j_k l_k}$ denotes the reaction rate function corresponding to the kth reaction.

Considering the special case of mass action networks, the form of the kinetics is $g_{il}(x) = \prod_{k=1}^{n} x^{\alpha_{ik}}$. However, in biological applications the reaction rates are often not limited to mass action kinetics, but we assume that they can be sorted to a finite set of biochemical kinetics $\mathcal{G}_i = \{G_1, G_2, \dots, G_{d_i}\}$ in the case of each complex C_i. Here d_i is the cardinality of $\mathcal{G}_i$, for $i = 1, \dots, m$. Each of these kinetics defines a relationship among the species of the complex, for example $G_1 =$ "Mass action kinetics," $G_2 =$ "Michaelis-Menten kinetics," $G_3 =$ "Hill kinetics," etc. With this notation $g_{il}(x)$ is associated with the kinetics G_l of the complex C_i.

In most kinetic biochemical ODE models the reaction rate functions have polynomial or rational function form. Thus we assume that the reaction rate can be written as a ratio of two terms as

$$g_{il}(x) = \frac{\Psi_i(x)}{D_{il}(x)} \tag{2.54}$$

where $\Psi_i(x)$ is a monomial function ($\Psi_i(x) = \prod_k x_k^{v_{ki}}$) and $D_{il}(x)$ is a positive polynomial function of the concentration vector (i.e., it has nonnegative coefficients and a positive zero-order term). To make the decomposition (2.52) unique, $g_{il}(x)$ must not contain any linear scaling constant and thus we fix the zero-order term in the denominator polynomial to 1, that is, $D_{il}(x)$ is written as $D_{il}(x) = 1 + \sum \alpha_{m_1, m_2, \dots, m_n} x_1^{m_1}, x_2^{m_2}, \dots, x_n^{m_n}$ where $\alpha \in \mathbb{R}_{0,+}$ and $m_1, m_2, \dots, m_n$ are nonnegative integers.

Since many of the kinetic biochemical ODE models can be represented in this form, we call these models biochemical reaction networks, or shortly, *bio-CRNs* [25]. Then we can characterize the bio-CRNs with the following four sets:

1. $\mathcal{S} = \{X_1, \ldots, X_n\}$ is the set of species or chemical substances.
2. $\mathcal{C} = \{C_1, \ldots, C_m\}$ is the set of complexes.
3. $\mathcal{G} = \cup_{i=1}^{m} \mathcal{G}_i$ the set of reaction rates (kinetics).
4. The set of biochemical reactions is

$$\mathcal{R} = \{(C_i, C_j, G_l) \mid C_i, C_j \in \mathcal{C}, G_l \in \mathcal{G}_i \text{ and } C_i \text{ is transformed to}$$
$$C_j \text{ with the kinetics } G_l\}$$

The set of species, complexes, and reactions with the kinetics uniquely determines the biochemical reaction network, which is denoted by $\Sigma = (\mathcal{S}, \mathcal{C}, \mathcal{G}, \mathcal{R})$.

2.3.4.1 Reaction Graph

The set of complexes together with the set of reactions give rise to the following directed, weighted graph representation of the bio-CRNs. The reaction graph $D = (V_d; E_d)$ consists of a finite nonempty set V_d of vertices (nodes), which represent the complexes $V_d = \{C_1, \ldots, C_m\}$, and a finite set of directed edges E_d, which represent the reactions. The edges are defined by triplets of the form $e_{(i,j,l)} = (C_i, C_j, G_l)$ for $i, j = 1, \ldots, m, i \neq j$, $l = 1, \ldots, d_i$, where $i, j,$ and l are the indices of the source complex, product complex, and the kinetics of the reaction, respectively. Furthermore, the principal reaction rate coefficients and the reaction kinetics are also used as weights on the edges.

Note that the introduced description allows multiple directed edges with different kinetics converting C_i to C_j for any $i, j = 1, \ldots, m, i \neq j$, which is a significant difference from the reaction graph of mass action systems, where multiedges are naturally not allowed. However, similarly to mass action CRNs, bio-CRNs also cannot contain loop-edges.

2.3.4.2 Dynamical Equations of Bio-CRNs

In bio-CRNs the reactions are consuming species of the source complexes while producing species in the product complexes. The corresponding equations give rise to a dynamic ODE model, which describes the concentration trajectories of the species. These models have a special structure, which allows us to read out the structure of the reaction graph from the equations. Motivated by the ODE structure of CRNs with mass action kinetics [16], the dynamics of the bio-CRN is written in the following form

$$\dot{x} = Y \cdot A_k \cdot P(x) \cdot \Psi(x) \tag{2.55}$$

where $x \in \bar{\mathbb{R}}_+^n$ is the concentration vector of the species, $Y \in \mathbb{N}_0^{n \times m}$ is the complex composition matrix, and $\Psi \colon \bar{\mathbb{R}}_+^n \to \bar{\mathbb{R}}_+^m$ is a vector function such that each element is a monomial

$$\Psi_i(x) = \prod_{j=1}^{n} x_j^{Y_{ji}}, \quad i = 1, \ldots, m \tag{2.56}$$

The generalized Kirchhoff matrix $A_k \in \mathbb{R}^{m \times \kappa}$ stores the principal reaction rate coefficients and it is a matrix with zero column sums. When only one kinetics is associated to each complex, for example in mass action networks, A_k is a square matrix, such that $[A_k]_{ij} = k_{ji}$ $(i \neq j)$ is the reaction rate coefficient of the reaction from complex C_j to complex C_i, and each diagonal element is the negative column-sum of corresponding off-diagonal elements, that is, $[A_k]_{ii} = -\sum_{j=1, j \neq i}^{m} [A_k]_{ji}$.

In the general bio-CRN case, maybe more than one kinetics is associated with each complex. In this case let us denote the number of kinetics corresponding to complex C_i by d_i. Then $A_k \in \mathbb{R}^{m \times \kappa}$ (where the total number of kinetics is denoted by $\kappa = \sum_{i=1}^{m} d_i$) can be written as a block matrix composed of m blocks of size $m \times d_i$ as

$$A_k = \left[A_k^{(1)} \ldots A_k^{(i)} \ldots A_k^{(m)} \right] \tag{2.57}$$

The jth row of the block $A_k^{(i)}$ $(i \neq j)$ contains the d_i principal reaction rate coefficients of the reactions from the complex C_i to complex C_j with kinetics index $l \in \{1, \ldots, d_i\}$ (i.e., $[A_k^{(i)}]_{jl} = k_{ijl}$). Furthermore, the elements of the ith row contain the negative sum of the other column elements of the same column as follows: $[A_k^{(i)}]_{il} = -\sum_{j=1, j \neq i}^{m} [A_k^{(i)}]_{jl}$. Thus A_k is a column conservation matrix (the sum of each column is 0). Naturally, $[A_k^{(i)}]_{jl} = 0$ means that there is no reaction in the network from complex C_i to C_j with kinetics index l. Due to this construction, the generalized Kirchhoff matrix has a close relationship with the graph of the reaction network: the nonzero elements correspond to the edges of the reaction graph and the location of the nonzero elements together with the numerical values defines the weights of the edges of the graph. The locations of the nonzero elements will be referred to as the *structure of the matrix A_k*.

The kinetic weighting mapping P: $\mathbb{R}^n \rightarrow \mathbb{R}^{\kappa \times m}$ arranges the denominator terms of the reaction rate functions (cf. Eq. 2.52) into a matrix form as follows

$$P(x) = \begin{bmatrix} P^{(1)}(x) & 0 & \cdots & 0 \\ 0 & P^{(2)}(x) & \cdots & 0 \\ & & \vdots & \\ 0 & 0 & \cdots & P^{(m)}(x) \end{bmatrix} \tag{2.58}$$

Here each block $P^{(i)}$ (for $i = 1, \ldots, m$) is of size $d_i \times 1$ and contains the denominators of the kinetics $g_{i1}, g_{i2}, \ldots, g_{id_i}$, for example, $[P^{(i)}]_l = \frac{1}{D_{il}(x)}$. The value of P at any x will be called a kinetic weighting matrix. Note that for MAL-CRNs the matrix $P(x)$ is the m-dimensional identity matrix.

From the earlier, it is clear that the bio-CRN Σ can be equivalently characterized either by $\Sigma = (\mathcal{S}, \mathcal{C}, \mathcal{G}, \mathcal{R})$ or by the set $\Sigma = (Y, A_k, P)$. While the former can primarily be used for the analysis of the network, the latter is more suitable for computational purposes.

Example 7 (The simplest bio-CRN model of the simple irreversible Michaelis-Menten kinetics). Most often the reaction kinetic expressions of the enzyme catalytic reactions are written in a rational function form instead of the detailed kinetic expressions in Example 5

$$\frac{d[S]}{dt} = -\frac{d[P]}{dt} = -\frac{K_1[S]}{K_M + [S]} \tag{2.59}$$

that is obtained from the original equations of the simple irreversible Michaelis-Menten mechanism by model simplification (reduction). The previous kinetic equations are called *Michaelis-Menten kinetic equations*.

The generalized reaction graph [25] of the biochemical reaction Eq. (2.59) consists of only two vertexes (complexes) and an overall reaction as seen in Fig. 2.3.

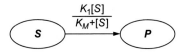

Fig. 2.3 The reaction graph of the bio-CRN model of the simple irreversible Michaelis-Menten kinetics.

2.3.4.3 Network Realization and Dynamical Equivalence

Given a set of ordinary differential equations, the realization problem is to find a possible bio-CRN with dynamics that is identical to the given equations. When the ODEs are given in the form

$$\dot{x} = Mg(x) \tag{2.60}$$

where $M \in \mathbb{R}^{n \times \kappa}$ is a coefficient matrix and $g: \mathbb{R}^n \to \mathbb{R}^\kappa$ is a kinetic vector function with each of its elements being in the rational form (2.54). Gábor et al. [25] provided necessary and sufficient conditions for the existence of such a network $\Sigma = (Y, A_k, P)$. If these conditions are fulfilled, we can write the right-hand side of Eq. (2.60) as

$$Mg(x) = YA_k P(x)\Psi(x) \quad \text{for all } x \in \mathbb{R}^n_+ \tag{2.61}$$

where $M = Y \cdot A_k$ and $g(x) = P(x)\Psi(x)$. Moreover, Y, A_k, P, and Ψ have the properties described in Section 2.3.4.2. If Eq. (2.61) holds, then $\Sigma = (Y, A_k, P)$ is called a *dynamically equivalent realization* of the *kinetic system* (2.60).

Similarly to the mass action case, it is easy to show that the realization problem does not have a unique solution in general, but one can find many possible different structures (simple examples can be found in [25]). Therefore, we say that the bio-CRNs $\Sigma' = (Y', A'_k, P')$ and $\Sigma'' = (Y'', A''_k, P'')$ are *dynamically equivalent*, if they give rise to the same dynamic equations of the form (2.60), that is,

$$Mg(x) = Y'A'_k P'(x)\Psi'(x) = Y''A''_k P''(x)\Psi''(x) \quad \text{for } \forall x \in \overline{\mathbb{R}}^n_+ \tag{2.62}$$

where Y', Y'' are nonnegative integer type matrices, A'_k, A''_k are generalized Kirchhoff matrices, P', P'' are rate weighting functions, and Ψ', Ψ'' are computed from Y' and Y'', respectively, according to Eq. (2.56).

Many structural (graph) properties of the network are realization dependent, thus it is of significant interest to find dynamically equivalent realizations with given properties (e.g., reversible, complex balanced, or weakly reversible realization) as formulated in [25].

Example 8 (A simple bio-CRN realization example). Consider the following set of ordinary differential equations

$$\dot{x}_1 = -k_{21}x_1^3 - \frac{k_{22}x_1^3}{1 + K_2 x_1} + 3k_{11}x_2^3 + \frac{3k_{12}x_2^3}{1 + K_1 x_2}$$

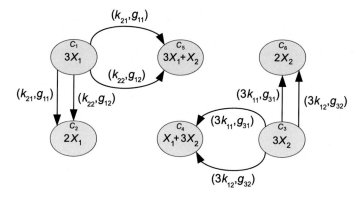

Fig. 2.4 Reaction graphs of the bio-CRN example from [25].

$$\dot{x}_2 = k_{21}x_1^3 + \frac{k_{22}x_1^3}{1 + K_2 x_1} - 3k_{11}x_2^3 - \frac{3k_{12}x_2^3}{1 + K_1 x_2}$$

It is easy to check that the necessary and sufficient conditions hold for this system.

The realization procedure presented in [25] created the complexes $\mathcal{C} = \{3X_1, 2X_1, 3X_2, X_1 + 3X_2, 3X_1 + X_2, 3X_2\}$ and the kinetics

$$g_{11} = x_1^3 \quad g_{12} = \frac{x_1^3}{1 + K_2 x_1}$$

$$g_{31} = x_2^3 \quad g_{32} = \frac{x_2^3}{1 + K_1 x_2}$$

The corresponding network can be seen in Fig. 2.4.

2.3.5 Extension With Input Term

An autonomous kinetic positive polynomial system corresponds to a dynamics under closed conditions. If one opens the system, that is, allows in- and/or outflows connecting the system and its environment then a special complex, the so-called zero complex is used in the description. Formally the zero complex corresponds to a zero column in the complex composition matrix Y, and positive constant terms in the kinetic ODE (2.33).

The usual and physically meaningful way is then to consider the inlet specie concentrations as manipulable input variables (see e.g., [27] for a detailed discussion). This leads to a state equation form with *linear input structure* as

$$\dot{x} = M \cdot \psi(x) + B \cdot u \tag{2.63}$$

where $x \in \mathbb{R}^n$ is the state (concentration) vector, $M \in \mathbb{R}^{n \times m}$ is the kinetic coefficient matrix, $\psi \in \mathbb{R}^n \to \mathbb{R}^m$ contains the reaction monomials of the open-loop system (as in Eq. 2.33), while $u \in \mathbb{R}^r$ is the input, and $B \in \mathbb{R}^{n \times r}$ is a constant matrix.

This structure will be used later in Section 5.2 to design stabilizing feedback controllers to polynomial systems.

2.4 BASIC RELATIONS BETWEEN KINETIC AND QP MODELS

A simple Example 4 in Section 2.2 has already indicated that an autonomous polynomial system can be recast into a QP model form by simple algebraic manipulations, if one restricts the domain of the state variables in the nonnegative orthant $\overline{\mathbb{R}}_+^n$.

2.4.1 Representing Kinetic Models With Mass Action Reaction Rates as QP Models

Because kinetic models with mass action reaction rates are positive polynomial systems by construction, the same algebraic method as in the simple Example 4 can be applied for them, too. The paper [6] gives a detailed description on how to represent kinetic systems with mass action reaction rates in QP form.

Let us assume mass action type kinetic systems that is described in the following form

$$\dot{x} = M \cdot \psi(x), \qquad \psi_j(x) = \prod_{i=1}^{n} x_i^{\alpha_{ji}}, \quad j = 1, \ldots, m \qquad (2.64)$$

with $x \in \overline{\mathbb{R}}_+^n$. As we have already seen in Section 2.3, this model is nonnegative. In order to bring the ℓth equation of Eq. (2.64) into a QP form of Eq. (2.12), we introduced the quasimonomials

$$p_{\ell_j} = \prod_{i=1}^{n} (x_i^{\alpha_{ji}} \cdot x_\ell^{-1}), \quad j = 1, \ldots, m$$

and formed a QP model using them.

2.4.2 LV Models as Kinetic Systems

In order to perform the reverse manipulation, that is, to construct a kinetic system model from an LV one, let us consider the ith equation of an

LV model (2.17) that is a positive polynomial system with second-order polynomial right-hand sides

$$\frac{dx_i}{dt} = x_i \left(\Lambda_i + \sum_{j=1}^m \mathcal{M}_{ij} x_j \right) = \Lambda_i x_i + \sum_{j=1}^m \mathcal{M}_{ij} x_j x_i = f_i(x) \qquad (2.65)$$

As we have already seen in Section 2.3, the necessary and sufficient condition of a polynomial model to have a kinetic realization is that the right-hand side function f_i of Eq. (2.48) is in the form $f_i(x) = -x_i g_i(x) + h_i(x)$, where g_i and h_i are polynomials with nonnegative coefficients. This is always fulfilled for the model (2.65) because of the multiplier x_i in its right-hand side. Indeed, the polynomial coefficients are the LV model parameters Λ_i and $\mathcal{M}_{ij}$ (for $i,j = 1,\ldots,m$). Now we can collect the first- or second-order terms with nonnegative coefficients to form the polynomial $h_i(x) = 0$, while the terms with negative coefficients will form the polynomial $-g_i(x)$. *Therefore, the necessary and sufficient condition for an LV model to have a kinetic realization is always fulfilled, and thus the canonical realization algorithm in [21] will produce a CRN realization with mass action kinetics.*

Other methods are also available to determine realizations of a kinetic polynomial system that will be described in Section 4.5.

A simple example of a CRN model of a simple LV system is given following.

Example 9 (CRN model of a simple LV model). Let us consider a simple 2D LV model in the following form

$$\begin{aligned} \frac{dx_1}{dt} &= x_1(\Lambda_1 + a_1 x_2) = \Lambda_1 x_1 + a_1 x_1 x_2 \\ \frac{dx_2}{dt} &= x_2(\Lambda_2 + a_2 x_1) = \Lambda_2 x_2 + a_2 x_1 x_2 \end{aligned} \qquad (2.66)$$

with $\Lambda_1 < 0$, $\Lambda_2 < 0$, and $a_2 > 0$.

Let us choose the complex composition matrix Y to be

$$Y = \begin{bmatrix} 1 & 0 & 1 & 0 \\ 0 & 0 & 1 & 0 \end{bmatrix}$$

Then there is a CRN kinetic realization of the previous model with the Kirchhoff matrix

$$A_k = \begin{bmatrix} -(k_1 + k_{10}) & 0 & 1 & 0 \\ 0 & -k_{20} & k_2 & 0 \\ k_1 & 0 & -k_2 & 0 \\ k_{10} & k_{20} & 0 & 0 \end{bmatrix}$$

where $k_{10}, k_{20}, k_1, k_2 > 0$, and

$$\Lambda_1 = -(k_1 + k_{10}), \quad a_1 = -k_2 + k_1, \quad \Lambda_2 = -k_{20}, \quad a_2 = k_2$$

The reaction graph of the previous CRN kinetic realization is shown in Fig. 2.5.

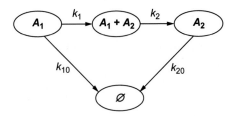

Fig. 2.5 *The reaction graph of the CRN realization of the simple LV model* (2.66).

CHAPTER 3

Model Transformations and Equivalence Classes

In this chapter, we consider closed positive polynomial systems that can be written in the form of an autonomous ODE

$$\frac{dx}{dt} = \mathcal{P}(x; k) \tag{3.1}$$

where $x \in \overline{\mathbb{R}}_+^n$ is the state variable, k is the vector of parameters, and $\mathcal{P}$ is a polynomial function.

Analysis and Control of Polynomial Dynamic Models with Biological Applications.
https://doi.org/10.1016/B978-0-12-815495-3.00003-1

Two subclasses of the previous systems that have been introduced in Chapter 2 are of interest:

- Quasipolynomial systems (QP systems) that are in the form

$$\frac{dx_i}{dt} = x_i \left(\lambda_i + \sum_{j=1}^{m} A_{i,j} \prod_{k=1}^{n} x_k^{B_{j,k}} \right), \quad i = 1, \ldots, n \qquad (3.2)$$

where $p_j = \prod_{k=1}^{n} x_k^{B_{j,k}}$ for $j = 1, \ldots, m$ are the quasimonomials.
- Chemical reaction networks (CRNs)

$$\frac{dx}{dt} = Y A_k \varphi(x), \qquad \varphi_j(x) = \prod_{k=1}^{n} x_k^{Y_{kj}}, \quad j = 1, \ldots, m \qquad (3.3)$$

3.1 AFFINE AND LINEAR POSITIVE DIAGONAL TRANSFORMATIONS

The simplest case for transforming a positive polynomial system is to apply an affine transformation to its state variables that correspond to rescaling and shifting the variables in an engineering sense. It is easy to see that this transformation will not emerge from the polynomial system class, but it may affect the positivity of the system.

3.1.1 Affine Transformations and Their Special Cases for Positive Polynomial Systems

Consider a positive polynomial system (3.1) and apply an arbitrary nonsingular transformation matrix $T \in \mathbb{R}^{n \times n}$ to its state variables (i.e., $x' = Tx$), that is, apply a *linear state transformation*. The transformed system will again be a polynomial system in the form

$$\frac{dx'}{dt} = TP(T^{-1}x'; k). \qquad (3.4)$$

One can apply a so-called *translation transformation* to Eq. (3.1) by introducing a new state variable $x' = x + \tau$ where $\tau \in \mathbb{R}^n$ is the translation parameter. By composing the two previous linear state transformations, the *affine transformation* is obtained, generated by the new state variable $x' = Tx + \tau$. *It is important to note that affine transformations preserve the topology of the phase space as well as the polynomial degree of the functions in the model* [28].

At the same time, affine transformations will generally lead out both from the subclass of QP systems and that of CRNs, unless a diagonal transformation matrix $T = D = \text{diag}\{d\}$, $d_i > 0, i = 1, \ldots, n$ and no translation (i.e., $\tau = 0$ is applied). Then a linear diagonal state transformation $x' = Dx$ is considered. We say that *both the QP and CRN subclasses are form-invariant with respect to the linear state transformation with a positive diagonal transformation matrix.*

3.1.2 Positive Diagonal Transformation of QP Systems

It is easy to see that neither the translation transformation nor the linear state transformation with an arbitrary nondiagonal transformation matrix preserves the QP form of an ODE. However, a *positive diagonal transformation* that is a special case of the affine transformation earlier with a diagonal transformation matrix and no translation is a subclass invariant transformation of QP systems.

The transformation parameter is the diagonal matrix $D = \text{diag}\{d\}$ where the elements of the vector $d \in \mathbb{R}^n$ are positive (i.e., $d_i > 0, i = 1, \ldots, n$). The state variables of the original QP model (3.2) are transformed as

$$x' = Dx, \quad x_i' = d_i x_i \tag{3.5}$$

The transformed quasimonomials are in the form

$$p_j' = \prod_{k=1}^{n} x'^{\mathcal{B}_{j,k}}_k = \left(\prod_{k=1}^{n} d_k^{-\mathcal{B}_{j,k}} \right) p_j = q(d)_j p_j, \quad j = 1, \ldots, m \tag{3.6}$$

The transformed model parameters can easily be obtained from the transformation formula of quasimonomials (Eq. 3.6). It shows that only the coefficient matrix $\mathcal{A}$ changes, that is,

$$\lambda' = \lambda, \quad \mathcal{A}' = \text{diag}\{q(d)\}\mathcal{A}, \quad \mathcal{B}' = \mathcal{B} \tag{3.7}$$

where $q(d) \in \overline{\mathbb{R}}_+^m$ is an element-wise positive vector that gives rise to a positive diagonal matrix $Q = \text{diag}\{q(d)\}$.

3.1.3 Positive Diagonal Transformation of CRNs: Linear Conjugacy

In [29] the authors define the concept of *linear conjugacy* through positive diagonal variable transformation. In their framework, *two CRNs denoted by Σ and Σ' are said to be linearly conjugate if there is a positive diagonal linear mapping which takes the flow of one network to the other.*

This means that we apply the positive diagonal transformation (3.5) to the state (concentration) variable of the CRN. It is known from [30] that such a transformation preserves the kinetic property of the system (2.46). It is easy to see that the transformed models share the same qualitative dynamics (e.g., number and stability of equilibria, persistence/extinction of species, dimensions of invariant spaces, etc.).

3.1.3.1 Linear Conjugacy of Networks With Mass Action Kinetics

Let $D \in \mathbb{R}^{n \times n}$ be a positive definite diagonal matrix. Using the same notations as in Eq. (3.5), the state transformation is performed as follows:

$$x' = D \cdot x, \quad x = D^{-1} \cdot x' \tag{3.8}$$

Applying it to the polynomial system (2.46) we get

$$\dot{x}' = D \cdot \dot{x} = DM \cdot \varphi(x) = DM \cdot \varphi(D^{-1}x') = DM\Phi_{D^{-1}} \cdot \varphi(x') \tag{3.9}$$

where $\Phi_{D^{-1}} \in \mathbb{R}^{n \times n}$ is a positive definite diagonal matrix so that $[\Phi_{D^{-1}}]_{ii} = \varphi_i(D^{-1} \cdot \underline{1})$ for $i \in \{1, \ldots, n\}$, φ_i is the ith coordinate function of φ, and $\underline{1} \in \mathbb{R}^n$ is a column vector with all coordinates equal to 1.

If we now consider CRNs with mass action kinetics, then we are interested whether the right-hand side of the transformed model in Eq. (3.9) can be written as a kinetic system in the form of Eq. (2.47). We say that a CRN realization given by (Y, A'_k) is a *linearly conjugate realization* of the kinetic system $\dot{x} = M\varphi(x)$ if there exists a positive definite diagonal matrix $D \in \mathbb{R}^{n \times n}$ such that

$$Y \cdot A'_k \cdot \psi(x) = D \cdot M \cdot \Phi_{D^{-1}} \cdot \varphi(x) \tag{3.10}$$

where $Y \in \mathbb{N}^{n \times m}$, $\psi : \mathbb{R}^n_+ \to \mathbb{R}^m_+$ with $\psi_j(x) = \prod_{i=1}^n x_i^{[Y]_{ij}}$ for $j \in \{1, \ldots, m\}$, and $A'_k \in \mathbb{R}^{m \times m}$ is a Kirchhoff matrix.

It is clear that dynamical equivalence is a special case of linear conjugacy, when the matrix D, and therefore the matrices D^{-1} and $\Phi_{D^{-1}}$ as well, are unit matrices.

Note that the monomial functions φ and ψ in Eqs. (2.47), (3.10) might be different. By applying the method described in [21], a suitable set of complexes corresponding to the canonical realization can be determined, but there might exist other possible complex sets as well. It is easy to see that the complexes determined by the monomials of function φ must be in the set $\mathcal{C}$, but arbitrary further complexes, which appear in the original kinetic equations with zero coefficients might be involved as well. These additional complexes modify the matrices Y and A'_k; therefore, we have to change the matrices M and Φ_D as well, in order to get the following equation:

$$Y \cdot A'_k \cdot \psi(x) = D \cdot M' \cdot \Phi'_{D-1} \cdot \psi(x) \qquad (3.11)$$

where the matrices $M' \in \mathbb{R}^{n \times m}$ and $\Phi'_{D-1} \in \mathbb{R}^{m \times m}$ have the same columns and diagonal entries as M and Φ_{D-1} belonging to the complexes determined by φ, and zero columns and diagonal entries equal to one belonging to all additional complexes, respectively.

Since there is the same monomial-type vector-mapping ψ on both sides of Eq. (3.11), the equation can be fulfilled if and only if the coefficients belonging to the same monomials are pairwise identical. This means that by using the notation $A_k = A'_k \cdot \Phi'^{-1}_{D-1}$, we can rewrite Eq. (3.11) as

$$Y \cdot A_k = D \cdot M' \qquad (3.12)$$

where A_k is a Kirchhoff matrix, too, obtained by scaling the columns of A'_k by positive constants. It is easy to see that this operation preserves the set of reactions, but changes the values of the nonzero entries. The actual reaction rate coefficients of the linearly conjugate network are contained in the matrix

$$A'_k = A_k \cdot \Phi'_{D-1} \qquad (3.13)$$

From now on, we will consider only linearly conjugate realizations on a fixed set of complexes. Therefore, we can assume that the kinetic ODEs are written in such a form that $\psi = \varphi$ and $M' = M$. A reaction network which is a linearly conjugate realization of a kinetic system can be identified by its matrices D, Y, and A'_k. However, since matrix Y is assumed to be fixed, we can simply denote this realization by the matrix pair (D, A_k).

3.1.3.2 Linear Conjugacy of CRNs With Rational Reaction Rates
As it is shown in [26], it is not difficult to extend the notion of linear conjugacy to bio-CRNs. First, we need three simple lemmas to easily handle the diagonal transformation. (The proofs can be found in [26].)

Lemma 1. *Let $A_k \in \mathbb{R}^{m \times \kappa}$ be a generalized Kirchhoff matrix corresponding to a network of m complexes and $d_1, d_2, \ldots, d_m$ kinetics ($\kappa = \sum_{i=1}^{m} d_i$) which belong to the complexes. Further on, let H be a positive diagonal matrix $H = diag(c)$, where $c \in \mathbb{R}^{\kappa}_+$. Then, $A'_k = A_k \cdot H$ is also a generalized Kirchhoff matrix with the same structure (same locations of the nonzero elements) as A_k.*

Lemma 2. *Let $P\colon \mathbb{R}^n \to \mathbb{R}^{\kappa \times m}$ be a kinetic weighting mapping corresponding to a network of m complexes and $d_1, d_2, \ldots, d_m$ kinetics ($\kappa = \sum_{i=1}^{m} d_i$) corresponding to the complexes. Furthermore, let T be a positive diagonal matrix $T = diag(c)$, where $c \in \mathbb{R}_+^n$. Then $\hat{P}(x) = P(Tx)$ is also a kinetic weighting mapping with the same structure as P.*

Lemma 3. *Let $P\colon \mathbb{R}^n \to \mathbb{R}^{\kappa \times m}$ be a kinetic weighting function corresponding to a network of m complexes and $d_1, d_2, \ldots, d_m$ kinetics ($\kappa = \sum_{i=1}^{m} d_i$) belonging to the complexes. Let S be a positive diagonal matrix $S = diag(s)$, where $s \in \mathbb{R}_+^m$ and let the block-diagonal matrix H be constructed as*

$$
H = \begin{pmatrix}
\left[diag\left(\frac{1}{s_1}\right) \right]_{d_1} & 0_{d_1 \times d_2} & \cdots \\
0_{d_2 \times d_1} & \left[diag\left(\frac{1}{s_2}\right) \right]_{d_2} & \cdots \\
\vdots & \ddots & \cdots \\
0_{d_m \times d_1} & \cdots & \left[diag\left(\frac{1}{s_m}\right) \right]_{d_m}
\end{pmatrix}
\tag{3.14}
$$

where $\left[diag\left(\frac{1}{s_i}\right) \right]_{d_i}$ denotes a diagonal block of size $d_i \times d_i$, which contains the constant $\frac{1}{s_i}$ in its diagonal, and $0_{d_i \times d_j}$ denotes a block of zeros of size $d_i \times d_j$. As earlier, d_i is the number of kinetics corresponding to complex C_i. Then

$$
P(x)S = H^{-1}P(x), \quad i.e., \ P(x) = HP(x)S \ for \ all \ x \in \mathbb{R}_+^n
\tag{3.15}
$$

Now we can state the following result which is the extension for the biochemical reaction networks of the linear conjugacy theorem presented by Johnston et al. [31, Theorem 2].

Consider a kinetic system of the form (2.60) with $M = Y \cdot A_k$ and $g(x) = P(x) \cdot \Psi(x)$, where $Y \in \mathbb{R}^{n \times m}$ is a complex composition matrix, $A_k \in \mathbb{R}^{m \times \kappa}$ is a generalized Kirchhoff matrix, $P(x) \in \mathbb{R}^{\kappa \times m}$ is a kinetic weighting matrix, and $\Psi_j(x) = \prod_{i=1}^{n} x_i^{Y_{ij}}$ for $j = 1, \ldots, m$. Assume that there exists an $n \times n$ positive definite diagonal matrix D such that

$$
Y \cdot A_b = D \cdot M
\tag{3.16}
$$

where $A_b \in \mathbb{R}^{m \times \kappa}$ is a generalized Kirchhoff matrix. Then the realization $(Y, \bar{A}_k, \bar{P})$ is linearly conjugate to Eq. (2.60) with

$$
\bar{A}_k = A_b \cdot \Phi_{D^{-1}}
\tag{3.17}
$$

$$
\bar{P}(x) = P(D^{-1}x)
\tag{3.18}
$$

and

$$\bar{\Phi}_{D^{-1}} = \begin{pmatrix} [\text{diag}(\Psi_1(c))]_{d_1} & 0_{d_1 \times d_2} & \cdots \\ 0_{d_2 \times d_1} & [\text{diag}(\Psi_2(c))]_{d_2} & \cdots \\ \vdots & & \ddots & \cdots \\ 0_{d_m \times d_1} & \cdots & & [\text{diag}(\Psi_m(c))]_{d_m} \end{pmatrix} \in \mathbb{R}^{\kappa \times \kappa}$$

(3.19)

where $[\text{diag}(\Psi_i(c))]_{d_i}$ denotes a diagonal block of size $d_i \times d_i$, which contains the constant $\Psi_i(c)$ in its diagonal, $0_{d_i \times d_j}$ denotes a block of zeros of size $d_i \times d_j$, and c is the vector containing diagonal elements of D^{-1}. As before, d_i denotes the number of kinetics corresponding to complex C_i.

It is worth showing the computations proving the previous result. Let us transform the solution of Eq. (2.60) using a positive diagonal transformation according to Eq. (3.8). Then we can write

$$\dot{x}' = D\dot{x} = D \cdot M \cdot P(x) \cdot \Psi(x) = D \cdot M \cdot P(D^{-1}x') \cdot \Psi(D^{-1}x')$$
$$= D \cdot M \cdot P(D^{-1}x') \cdot \Phi_{D^{-1}} \cdot \Psi(x') = D \cdot M \cdot \bar{P}(x') \cdot \Phi_{D^{-1}} \cdot \Psi(x')$$

(3.20)

where $\Phi_{D^{-1}}$ is a diagonal matrix given by

$$\Phi_{D^{-1}} = \text{diag}(\Psi(D^{-1} \cdot \mathbf{1}^n)) = \text{diag}(\Psi(c))$$ (3.21)

and $\mathbf{1}^n$ denotes the n-dimensional vector with all coordinates equal to 1. According to Lemma 2, $\bar{P}(x')$ is a kinetic weighting matrix, and therefore, we can apply Lemma 3 as

$$\bar{P}(x') \cdot \Phi_{D^{-1}} = \bar{\Phi}_{D^{-1}} \cdot \bar{P}(x')$$ (3.22)

where $\bar{\Phi}_{D^{-1}}$ is given by Eq. (3.19). Therefore, using Eq. (3.22) we can write Eq. (3.20) as

$$\dot{x}' = D \cdot M \cdot \bar{\Phi}_{D^{-1}} \cdot \bar{P}(x') \cdot \Psi(x')$$ (3.23)

Clearly, if Eq. (3.16) holds, then

$$\dot{x}' = Y \cdot A_b \cdot \bar{\Phi}_{D^{-1}} \cdot \bar{P}(x') \cdot \Psi(x')$$ (3.24)

that is a kinetic system characterized by $(Y, \bar{A}_k, \bar{P})$, since, according to Lemma 1, $\bar{A}_k$ given in Eq. (3.17) is a generalized Kirchhoff matrix.

3.2 NONLINEAR DIAGONAL TRANSFORMATIONS

If one considers natural nonlinear transformations for a positive polynomial system (3.1) that do not emerge from this system class, then three overlapping transformation categories can be identified: (i) nonlinear state variable transformations, (ii) phase-space transformations, and (iii) time-rescaling. Although polynomial nonlinear state variable transformations do not emerge from the class of positive polynomial system, but they are not invariant for the QP system and CRN subclass in the general case, we restrict our attention to the phase-space and time-rescaling transformations here.

3.2.1 X-Factorable Transformation

The so-called *translated X-factorable transformation* was defined in [18] to allow us to derive nonlinear chemical kinetic schemes (i.e., CRN models) from mechanical and electrical dynamical systems. This transformation is applied to only the nonlinear mapping $\mathcal{P}$ of the polynomial ODE model (3.1); therefore, it is called a *phase-space transformation*.

In the general case, the translated X-factorable transformation can be applied to any ODE with a smooth right-hand side function f in the form

$$\frac{dx}{dt} = f(x) \tag{3.25}$$

with $x \in \mathbb{R}^n$. Let us introduce the matrix function $X' = \text{diag}\{x'\} \in \mathbb{R}^{n \times n}$, and the translation parameter $\tau \in \mathbb{R}^n$, then the transformed version of Eq. (3.1) is as follows

$$\frac{dx'}{dt} = X'f(x' - \tau) \tag{3.26}$$

The transformation can move the singular solutions of Eq. (3.25) to any desired point in $\mathbb{R}^n$; thus, it may move all the steady-state points into the positive orthant. It is important to note that the topological properties of a positive steady-state point of the translated ODE $\frac{dx}{dt} = f(x - \tau)$ are identical to that of Eq. (3.26) as long as the Jacobians of Eq. (3.26) evaluated at all positive singular solutions have eigenvalues that are sign-wise identical to the Jacobian eigenvalues of the translated ODE evaluated at all of its singularities.

At the same time, this nonlinear transformation has an effect on dynamical behavior. Due to nonlinearity, a substantial compression of trajectories occurs close to the boundary of the positive orthant. This distortion is weak or negligible far from the boundary. Therefore, if the singular points of the translated ODE are far from the boundary, then the behavior of the translated X-factorable transformed ODE (3.26) is approximately dynamically equivalent to the translated one.

In the case of *strictly positive polynomial systems*, however, there is no need to translate Eq. (3.1) (i.e., $\tau = 0$), as every singularity is strictly positive. In this case *the X-factorable transformation results in a positive polynomial system, too*. In addition, *the X-factorable transformed version of a positive polynomial system* (3.1)

$$\frac{dx}{dt} = X\mathcal{P}(x; k) \tag{3.27}$$

becomes kinetic; therefore, a CRN realization will always exist.

3.2.2 State-Dependent Time-Rescaling

Another important transformation is the state-dependent time-rescaling or time reparametrization transformation that applies to any positive polynomial system.

The transformation is applied to the only independent variable of the model (i.e., to the time t). Let $\omega = [\omega_1, \ldots, \omega_n]^T \in \mathbb{R}^n$, and let us introduce a new time, t' such that

$$dt = \prod_{k=1}^{n} x_k^{\omega_k} dt' \tag{3.28}$$

It is easy to see that this transformation will also result in a positive polynomial system, but the polynomial degrees will change in the general case.

3.2.2.1 Time-Rescaling Transformation of QP Models

It is shown, for example, in [14] that the state-dependent time-rescaling (or reparametrization) transforms the original QP system into the following (also QP) form:

$$\frac{dx_i}{dt'} = x_i \sum_{j=1}^{m+1} \mathcal{A}'_{i,j} \prod_{k=1}^{n} x_k^{\mathcal{B}'_{j,k}}, \quad i = 1, \ldots, n \tag{3.29}$$

where $\mathcal{A}' \in \mathbb{R}^{n \times (m+1)}$, $\mathcal{B}' \in \mathbb{R}^{(m+1) \times n}$, and

$$\mathcal{A}'_{i,j} = \mathcal{A}_{i,j}, \quad i = 1, \ldots, n; \, j = 1, \ldots, m \tag{3.30}$$

$$\mathcal{A}'_{i,m+1} = \lambda_i, \quad i = 1, \ldots, n \tag{3.31}$$

and

$$\mathcal{B}'_{i,j} = \mathcal{B}_{i,j} + \omega_j, \quad i = 1, \dots, m; \ j = 1, \dots, n \tag{3.32}$$

$$\mathcal{B}'_{m+1,j} = \omega_j, \quad j = 1, \dots, n \tag{3.33}$$

It can be seen that the number of monomials is increased by 1 and vector λ' is generally 0 in the transformed system.

A special case of the time-rescaling (time reparametrization) transformation occurs when the following relation holds:

$$\omega^T = -b_j, \quad 1 \le j \le m \tag{3.34}$$

where b_j is an arbitrary row of the B matrix of the original system (3.2). From Eqs. (3.32), (3.33) we can see that in this case, the jth row of $\tilde{B}$ is a zero vector. This means that the number of monomials in the transformed system (3.29) remains the same as in the original QP system (3.2) and a nonzero $\tilde{\lambda}$ vector that is equal to the jth column of A appears in the transformed system (e.g., see [14]).

The most important *properties of the time-rescaling transformation* in the case of QP systems are as follows.

1. *Monomials.* The set of monomials $p'_1, \dots, p'_{m+1}$ for the rescaled system can be written in terms of the original monomials:

$$p'_j = \prod_{k=1}^{n} x_k^{\omega_k} \cdot \prod_{k=1}^{n} x_k^{B_{j,k}} = \prod_{k=1}^{n} x_k^{B_{j,k}+\omega_k}, \quad j = 1, \dots, m$$

and

$$p'_{m+1} = \prod_{k=1}^{n} x_k^{\omega_k}$$

or using a shorter notation:

$$p'_j = r \cdot p_j, \quad j = 1, \dots, m$$

$$p'_{m+1} = r$$

where p_j is the corresponding monomial in the original system, and

$$r = \prod_{k=1}^{n} x_k^{\omega_k}$$

2. *Equilibrium points.* Since the equations of the rescaled system (3.29) can be written as

$$\frac{dx_i}{dt'} = x_i \left(\lambda_i + \sum_{j=1}^{m} \mathcal{A}_{i,j} \prod_{k=1}^{n} x_k^{\mathcal{B}_{j,k}} \right) \prod_{k=1}^{n} x_k^{\omega_k}, \quad i = 1, \ldots, n \quad (3.35)$$

and we assume that $x_i > 0$, $i = 1, \ldots, n$, it is clear that the equilibrium point x^* of the original QP system (3.2) is also an equilibrium point of the rescaled system (3.35).

3. *Local stability.* Let us denote the Jacobian matrix of the original QP system (3.2) at the equilibrium point by $J(x^*)$. Then the Jacobian matrix of the time-rescaled QP system at the equilibrium point can be computed by using the formula described in [32]:

$$J'(x^*) = X^* \cdot \mathcal{A}' \cdot Z'^* \cdot \mathcal{B}' \cdot (X^*)^{-1} = r^* \cdot J(x^*) = \prod_{k=1}^{n} x_k^{* \ \omega_k} \cdot J(x^*) \quad (3.36)$$

where

$$Z'^* = \mathrm{diag}\{p_1^*, \ldots, p_m^*, p_{m+1}^*\}, \quad X^* = \mathrm{diag}\{x_1^*, \ldots, x_n^*\}$$

are the quasimonomials of the time-rescaled system and the system variables in the equilibrium point. From Eq. (3.36) one can see that (as we naturally expect) local stability is not affected by the time-rescaling, because this transformation just multiplies the eigenvalues of the Jacobian by a positive constant r^*.

4. *Global stability.* Rewriting Eq. (3.28) gives

$$\frac{dt}{dt'} = \prod_{k=1}^{n} (x_k(t'))^{\omega_k} \quad (3.37)$$

from which we can see that t is a strictly monotonously increasing continuous and invertible function of t'. This means that global stability of the QP system in the rescaled time t' is equivalent to global stability in the original time scale t.

3.3 QUASIMONOMIAL TRANSFORMATION AND THE CORRESPONDING EQUIVALENCE CLASSES OF QP SYSTEMS

In this section we make use of the *logarithmic form of QP models* (3.2) introduced in Section 2.2.1 as Eq. (2.13) by introducing the notation of element-wise logarithm (denoted by $\mathrm{Ln}(x)$) of a vector $x \in \mathbb{R}_+^n$. Then the logarithmic form of the QP model (3.2) becomes

$$\begin{aligned} \frac{d\mathrm{Ln}(x)}{dt} &= \lambda + \mathcal{A}p \\ \mathrm{Ln}(p) &= \mathcal{B}\mathrm{Ln}(x) \end{aligned} \quad (3.38)$$

3.3.1 Quasimonomial Transformation (QM Transformation)

Given a QP model with its parameters $(\mathcal{A}, \mathcal{B}, \lambda)$, the quasimonomial transformation (abbreviated as QM transformation) is defined as

$$x_i' = \prod_{k=1}^{n} x_k^{C_{ik}}, \quad i = 1, \ldots, n \tag{3.39}$$

where C is an arbitrary invertible matrix, the parameter of the transformation. Note that this transformation is an invertible state transformation that does not emerge from the QP model subclass, so *QP models are form-invariant with respect to it.*

With the previous transformation rules for the variables, we can easily deduce the *transformation rules of the QP model parameters,* which are

$$\mathcal{B}' = \mathcal{B} \cdot C, \quad \mathcal{A}' = C^{-1} \cdot \mathcal{A}, \quad \lambda' = C^{-1} \cdot \lambda \tag{3.40}$$

At the same time, the number of quasimonomials does not change with this transformation. It is important to note that the products $\mathcal{B} \cdot \mathcal{A} = \mathcal{B}' \cdot \mathcal{A}'$ and $\mathcal{B} \cdot \lambda = \mathcal{B}' \cdot \lambda'$ remain unchanged, too.

Therefore, the subclass of QP models is split into *classes of equivalence* [12] according to the values of the products $\mathcal{M} = \mathcal{B} \cdot \mathcal{A}$ and $\Lambda = \mathcal{B} \cdot \lambda$ which are *the invariants under QM transformation.*

3.3.2 The Lotka-Volterra (LV) Form and the Invariants

The time-derivative of the element-wise logarithm of the monomials $(\text{Ln}(p))$ can be easily derived using the logarithmic form (3.38) of QP models, from which the following set of nonlinear ODE is obtained in its logarithmic form:

$$\frac{d\text{Ln}(p)}{dt} = \Lambda + \mathcal{M}p \tag{3.41}$$

This form is called *the Lotka-Volterra (abbreviated as LV) canonical form of a QP model equivalence class.*

Note that this model is also a QP model with the special parameters $(\mathcal{A}_{\text{LV}} = \mathcal{M}, \mathcal{B}_{\text{LV}} = I, \lambda = \Lambda)$, where the invariants of the QM transformation $\mathcal{M}, \Lambda$ appear in the parameter set.

However, in the case of $m > n$ (i.e., more quasimonomials than state variables), the parameter matrix $\mathcal{M}$ of the LV canonical form is rank deficient, even in the case when rank $\mathcal{B} = n$ (full rank) [33]. In order to show this, let us introduce $(m - n)$ new variables $x_{n+1}, \ldots, x_m$ defined in such a way that $\frac{dx_k}{dt} = 0$, $k = n + 1, \ldots, m$. Then the extended parameters $(\tilde{\mathcal{A}}, \tilde{\mathcal{B}}, \tilde{\lambda})$ with the extended state vector $\tilde{x}$ are as follows:

$$
\tilde{x} = \begin{bmatrix} x_1 \\ \vdots \\ x_n \\ \hline 1 \\ \vdots \\ 1 \end{bmatrix}, \quad
\tilde{\lambda} = \begin{bmatrix} \lambda \\ \hline \emptyset \end{bmatrix}, \quad
\tilde{\mathcal{A}} = \begin{bmatrix} \mathcal{A} \\ \hline \emptyset \end{bmatrix}, \quad
\tilde{\mathcal{B}} = \begin{bmatrix} \mathcal{B}^* & \emptyset \\ \hline \mathcal{B}^+ & I \end{bmatrix}
$$

$$(3.42)$$

where $\emptyset$ is a zero block of appropriate dimension, $\mathcal{B}^* \in \mathbb{R}^{n \times n}$ is an invertible matrix, and I is the unit matrix of appropriate dimension. The previous extension of the state variables can be regarded as *an embedding of the state vector $x \in \mathbb{R}_+^n$ into a higher dimensional space of $\tilde{x} \in \mathbb{R}_+^m$ such that the dynamics will move in a lower dimensional (n-dimensional) manifold of $\mathbb{R}_+^m$.*

With this choice of the extension, the extended matrix $\tilde{\mathcal{B}}$ will be invertible, and can be used as a parameter of a QM transformation that results in the LV form of the original QP model. But the *maximal rank of the parameter matrix $\mathcal{M} = \tilde{\mathcal{B}} \cdot \tilde{\mathcal{A}}$ will only be n* (the maximal rank of $\tilde{\mathcal{A}}$). In addition, we can use the second equation of the logarithmic form (3.38) with the partitioned quasimonomial vector

$$
p = \begin{bmatrix} p_1 \\ \vdots \\ p_n \\ \hline p_{n+1} \\ \vdots \\ p_m \end{bmatrix} = \begin{bmatrix} p^* \\ \hline p^+ \end{bmatrix}
$$

to develop a set of nonlinear algebraic equations relating the quasimonomial partition p^* of the vector p to the other one (p^+) as follows

$$\mathrm{Ln}(p^+) = \mathcal{B}^+ (\mathcal{B}^*)^{-1} \mathrm{Ln}(q^*) = \mathcal{B}^{**} \mathrm{Ln}(q^*) \tag{3.43}$$

This way we obtained that *the LV form of a QP model equivalence class in the $m > n$ case is the LV model (3.41) with a rank-deficient coefficient matrix $\mathcal{M}$ constrained by the nonlinear algebraic equations (3.43).*

The previous special LV model explicitly contains the so-called *quasipolynomial invariants* of the LV model (3.41). More about the notion and computations of quasipolynomial invariants will be given in Section 4.3.2.

3.4 EMBEDDING TRANSFORMATIONS AND THE RELATIONSHIP BETWEEN CLASSES OF POSITIVE POLYNOMIAL SYSTEMS

The state and phase space transformations in this chapter enable us to transform certain models from a class (e.g., from the QP models) of the positive polynomial systems to another class (i.e., to the CRN models). Further possibilities exist with the introduction of the so-called *embedding transformations*. In the general case, an embedding transformation is a state transformation that introduces $s > 0$ new state variables to extend the original state space $\overline{\mathbb{R}}_+^n$ to have $\overline{\mathbb{R}}_+^m$ with $m = n+s$. This is done by adding s nonlinear algebraic relationships which relate the new variables to the other ones; therefore, the dynamics of the extended higher dimensional system evolves on an n-dimensional manifold of $\mathbb{R}_{>0}^m$.

An example of an embedding transformation was already given in Section 3.3.2, with the derivation of the LV form of QP models.

3.4.1 Embedding Smooth Nonlinear Models Into QP Form (QP Embedding)

A wide class of nonlinear autonomous systems with smooth nonlinearities can be embedded into QP model form if they satisfy *two requirements* [12].

1. The set of nonlinear ODEs should be in the form:

$$\frac{dx_s}{dt} = x_s \left(\sum_{i_{s1},\ldots,i_{sn},j_s} a_{i_{s1},\ldots,i_{sn}j_s} x_1^{i_{s1}}, \ldots, x_n^{i_{sn}} f(\overline{x})^{j_s} \right) \tag{3.44}$$

$$x_s(t_0) = x_s^0, \quad s = 1,\ldots,n$$

where $f(\overline{x})$ is some scalar valued function of a subset $\overline{x}$ of the state vector x, which *is not reducible to quasimonomial form* containing terms in the form of $\prod_{k=1}^n x_k^{\Gamma_{jk}}$, $j = 1,\ldots,m$ with Γ being a real matrix.
2. Furthermore, we require that the partial derivatives of the model (3.44) fulfill

$$\frac{\partial f}{\partial x_s} = \sum_{e_{s1},...,e_{sn},e_s} b_{e_{s1},...,e_{sn}e_s} x_1^{e_{s1}},\ldots,x_n^{e_{sn}} f(\overline{x})^{e_s}$$

The embedding is performed by introducing a new auxiliary variable

$$\eta = f^q \prod_{s=1}^{n} x_s^{p_s}, \quad q \neq 0 \tag{3.45}$$

Then, instead of the non-QP nonlinearity f, we can write the original set of Eq. (3.44) into QP form

$$\frac{dx_s}{dt} = x_s \left(\sum_{i_{s1},...,i_{sn},j_s} \left(a_{i_{s1},...,i_{sn}j_s} \eta^{j_s/q} \prod_{k=1}^{n} x_k^{i_{sk}-\delta_{sk}-j_s p_k/q} \right) \right), \quad s = 1,\ldots,n \tag{3.46}$$

where $\delta_{sk} = 1$ if $s = k$ and 0 otherwise. In addition, a new QP ODE appears for the new variable η:

$$\frac{d\eta}{dt} = \eta \left[\sum_{s=1}^{n} \left(p_s x_s^{-1} \frac{dx_s}{dt} \right. \right.$$

$$\left. \left. + \sum_{i_{s\alpha}j_s e_{s\alpha},e_s} a_{i_{s\alpha}j_s} b_{e_{s\alpha},e_s} q \cdot \eta^{(e_s+j_s-1)/q} \cdot \prod_{k=1}^{n} x_k^{i_{sk}+e_{sk}+(1-e_s-j_s)p_k/q} \right) \right]$$

$$\alpha = 1,\ldots,n \tag{3.47}$$

It is important to observe that the previous embedding is not unique, because we can choose the parameters p_s and q in Eq. (3.45) in many different ways; the simplest solution is to choose ($p_s = 0$, $s = 1,\ldots,n$; $q = 1$). Since the embedded QP system includes the original differential variables x_i, $i = 1,\ldots,n$, it is clear that the stability of the embedded system (3.46), (3.47) implies the stability of the original system (3.44).

Example 10 (QP embedding of a fermenter model). Consider a simple fermentation process with nonmonotonous reaction kinetics, described by the non-QP input-affine state-space model

$$\dot{x}_1 = \mu(x_2)x_1 + \frac{(X_F - x_1)F}{V}$$

$$\dot{x}_2 = -\frac{\mu(x_2)x_1}{Y} + \frac{(S_F - x_2)F}{V} \tag{3.48}$$

$$\mu(x_2) = \mu_{max} \frac{x_2}{K_2 x_2^2 + x_2 + K_1}$$

where the state variables x_1 and x_2 are the biomass and the substrate concentrations, respectively. The inlet substrate and biomass concentrations denoted by S_F and X_F are the manipulated inputs.

By introducing a new differential variable $\eta = \frac{1}{K_2 x_2^2 + x_2 + K_1}$, one arrives at a third-differential equation

$$\dot{\eta} = -\frac{2K_2 x_2 + 1}{(K_2 x_2^2 + x_2 + K_1)^2} \cdot \dot{x}_2 \tag{3.49}$$

that completes the ones for x_1 and x_2. Thus the original system (3.48) can be represented by three differential equations in input-affine QP form:

$$\dot{x}_1 = x_1 \left(\mu_{\max} x_2 \eta - \frac{F}{V} \right) + x_1 \left(x_1^{-1} \frac{F}{V} \right) X_F$$

$$\dot{x}_2 = x_2 \left(-\frac{\mu_{\max}}{Y} x_1 \eta - \frac{F}{V} \right) + x_2 \left(x_2^{-1} \frac{F}{V} \right) S_F \tag{3.50}$$

$$\dot{\eta} = \eta \left(\frac{2\mu_{\max} K_2}{Y} x_1 x_2^2 \eta^2 + \frac{2K_2 F}{V} x_2^2 \eta + \frac{\mu_{\max}}{Y} x_1 x_2 \eta^2 + \frac{F}{V} x_2 \eta \right)$$

$$+ \eta \left(-\frac{2K_2 F}{V} x_2 \eta - \frac{F}{V} \eta \right) S_F$$

3.4.2 Embedding Rational Functions Into Polynomial Form (CRN embedding)

Rational functions play a key role in the reaction rate expressions in biochemical reaction networks, as we have already seen in Section 2.3. There the *elementary biochemical reaction rate function* was given in the following form:

$$r = k \cdot g(x) = k \frac{M_i(x)}{D(x)} = kM(x) \cdot \frac{1}{D(x)} \tag{3.51}$$

where $M(x) = \prod_{l=1}^{n} x_l^{v_l} \geq 0$ is a monomial with the nonnegative integers v_{li}, and k is the *principal reaction rate coefficient*. The denominator function $D(x)$ is assumed to be a positive polynomial

$$D(x) = 1 + \sum \alpha_{m_1, m_2, \dots, m_n} x_1^{m_1}, x_2^{m_2}, \dots, x_n^{m_n} > 0 \tag{3.52}$$

where $\alpha \in \overline{\mathbb{R}}_+$ and $m_1, m_2, \dots, m_n$ are nonnegative integers.

Clearly, Eq. (3.51) contains a nonpolynomial factor $f(x) = \frac{1}{D(x)}$, and satisfies the two requirements of embedding earlier.

For the sake of simplicity, let us assume that the denominator is only a function of one of the state variables (i.e., $D(x) = D(x_s)$), and the reaction with the reaction rate (Eq. 3.51) consumes x_s. Then we can introduce the new auxiliary variable η in the simplest way

$$\eta = \frac{1}{D(x_s)} \tag{3.53}$$

Then the new differential equation becomes

$$\frac{d\eta}{dt} = -\eta^2 \cdot D'(x_s) \tag{3.54}$$

where $D'(x_s) = \frac{\partial D}{\partial x_s}$ is a positive polynomial, so this equation is clearly kinetic. At the same time, the original rate Eq. (3.51) becomes

$$r = kM(x) \cdot \eta \tag{3.55}$$

which also remains kinetic.

This way one can embed biochemical reaction rate functions being rational functions into CRNs obeying the mass action law.

3.4.3 Generality and Relationship Between Classes of Positive Polynomial Systems

The various transformations described in this chapter can be used to relate the different classes and subclasses of positive polynomial systems, and other smooth nonlinear systems to each other, as it is depicted in Fig. 3.1.

The class and subclass relations are clearly visible in the figure with the LV models being the most special subclasses of positive polynomial systems.

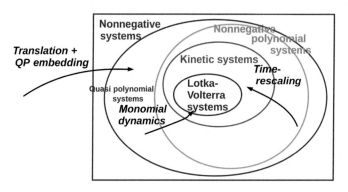

Fig. 3.1 The relationship between classes of positive polynomial systems induced by transformations and embeddings.

Translation and QP embedding can be used to map a wide class of smooth nonlinear systems into QP model form, that is the most wide class of positive polynomial systems. With this model form, one can use invertible state transformations to apply the structural analysis techniques available for CRN models, too.

CHAPTER 4

Model Analysis

Analysis and Control of Polynomial Dynamic Models with Biological Applications.
https://doi.org/10.1016/B978-0-12-815495-3.00004-3

4.1 STABILITY ANALYSIS OF QP MODELS

Stability is the most important dynamic property that is not only used for analysis but also for controller design. Stability analysis of general nonlinear systems is usually hard (see Section B.4.2), but one may utilize the special structure of a certain model class to obtain computationally feasible methods.

This section reformulates the time-decreasing condition of a class of Lyapunov functions for QP and Lotka-Volterra (LV) systems (see Section 2.2 for the introduction of the model forms of these systems) so that off-the-shelf numerical solvers can be used for their global stability analysis.

4.1.1 Local Stability Analysis of QP and LV Models

Here the (nontrivial) stationary solution x^* is supposed to be determined from the steady-state version of Eq. (2.12)

$$0 = x_i^* \left(\lambda_i + \sum_{j=1}^{m} A_{ij} \prod_{k=1}^{n} x_k^{*B_{jk}} \right) \tag{4.1}$$

$$i = 1, \ldots, n, \quad m \geq n$$

Let us denote the Jacobian matrix of the original QP system (2.12) at the equilibrium point by $J(x^*)$. The Jacobian matrix of the QP system at the equilibrium point can be computed by using the formula described in [32]:

$$J(x^*) = X^* \cdot A \cdot P^* \cdot B \cdot X^{*-1} \tag{4.2}$$

where X^* and Q are the following diagonal matrices:

$$X^* = \begin{bmatrix} x_1^* & 0 & \cdots & 0 \\ 0 & x_2^* & \ddots & \vdots \\ \vdots & \ddots & \ddots & 0 \\ 0 & \cdots & 0 & x_n^* \end{bmatrix} \quad P^* = \begin{bmatrix} p_1^* & 0 & \cdots & 0 \\ 0 & p_2^* & \ddots & \vdots \\ \vdots & \ddots & \ddots & 0 \\ 0 & \cdots & 0 & p_m^* \end{bmatrix}$$

4.1.1.1 Jacobian in the LV Case
In the classical LV case the Jacobian matrix can be simplified because LV systems are special QP systems, so the *Jacobian matrix for the LV system is* [34]:

$$J_{LV}(p^*) = P^* \cdot \mathcal{M} \qquad (4.3)$$

4.1.2 Global Stability Analysis Through the Solution of Linear Matrix Inequalities

Henceforth, it is assumed that x^* is a positive equilibrium point, that is $x^* \in \mathbb{R}^n_+$ in the QP case, and similarly, $p^* \in \mathbb{R}^m_+$ is a positive equilibrium point in the LV case. For LV systems, there is a well-known candidate Lyapunov function family [13, 14], which is in the form

$$V(p) = \sum_{i=1}^m c_i \left(p_i - p_i^* - p_i^* \ln \frac{p_i}{p_i^*} \right) \qquad (4.4)$$

$$c_i > 0, \quad i = 1 \dots m,$$

where $p^* = (p_1^*, \dots, p_m^*)^T$ is the equilibrium point corresponding to the equilibrium x^* of the original QP system (2.12). The time derivative of the Lyapunov function (4.4) is

$$\dot{V}(p) = \frac{1}{2}(p - p^*)(C\mathcal{M} + \mathcal{M}^T C)(p - p^*) \qquad (4.5)$$

where $C = \text{diag}\{c_1, \dots, c_m\}$ and $\mathcal{M}$ is the invariant characterizing the LV form (2.17). Therefore, the nonincreasing nature of the Lyapunov function is equivalent to a feasibility problem over the following set of *linear matrix inequality* (LMI) constraints:

$$\begin{aligned} C\mathcal{M} + \mathcal{M}^T C &\leq 0 \\ C &> 0 \end{aligned} \qquad (4.6)$$

where the unknown matrix is C, which is diagonal and contains the coefficients of Eq. (4.4).

Remark

1. Note the similarity of the stability conditions with continuous time LTI systems: for a system with state matrix A to be asymptotically stable, there must be positive definite matrices P and Q such that $A^T P + PA = -Q$ (the *Lyapunov equation*). If P is a diagonal matrix, A is said to be

diagonally stable [35]. So Eq. (4.6) requires the diagonal stability of the LV coefficient matrix $\mathcal{M}$. (More about matrix stability and special matrices can be found in Section B.2.)

2. It is important to mention that the strict positivity constraint on c_i can be somewhat relaxed in the following way [14]: if the equations of the model (2.12) are ordered in such a way that the first n rows of B are linearly independent, then $c_i > 0$ for $i = 1, \ldots, n$ and $c_j = 0$ for $j = n+1, \ldots, m$ still guarantees global stability.

3. It is examined and proved in [13, 14] that the global stability of Eq. (2.17) with Lyapunov function (4.4) implies the boundedness of solutions and global stability of the original QP system (2.12). It is stressed that *global stability is restricted to the positive orthant $\mathbb{R}_+^n$ only for QP and LV models*, because it is their original domain (see the definition in Eq. 2.12).

4. It is also important that the global stability of the equilibrium points of Eq. (2.12) with Lyapunov function (4.4) does not depend on the value of the vector L as long as the equilibrium points are in the positive orthant [14].

The possibilities to find a Lyapunov function that proves the global asymptotic stability of a QP system can be increased by using time-reparametrization [36], that is described in Section 4.1.3.

4.1.3 Extension With Time Reparametrization

As it was mentioned in Section 3.2.2, time-dependent rescaling or time reparametrization of states extends the possibilities of local and global stability analysis. At the same time, this transformation preserves the QP form, so it can be applied for QP systems.

4.1.3.1 The Time-Reparametrization Problem as a Bilinear Matrix Inequality

We denote an $n \times m$ matrix containing zero elements by $0^{n \times m}$. Let us define two auxiliary matrices by extending $\mathcal{A}$ with a zero column and $\mathcal{B}$ with a zero row, that is,

$$\bar{A} = \left[\mathcal{A} \mid 0^{n \times 1} \right] \in \mathbb{R}^{n \times (m+1)} \tag{4.7}$$

and

$$\bar{B} = \left[\begin{array}{c} \mathcal{B} \\ \hline 0^{1 \times n} \end{array} \right] \in \mathbb{R}^{(m+1) \times n} \tag{4.8}$$

Then $\tilde{A}$ and $\tilde{B}$ can be written as

$$\tilde{A} = [A|\lambda] = \bar{A} + [0^{n \times m}|\lambda] \tag{4.9}$$

and

$$\tilde{B} = \begin{bmatrix} \mathcal{B}_{1,.} + \omega^T \\ \mathcal{B}_{2,.} + \omega^T \\ \vdots \\ \mathcal{B}_{m,.} + \omega^T \\ \omega^T \end{bmatrix} = \bar{B} + S \cdot \Omega \tag{4.10}$$

where

$$\Omega = \mathrm{diag}(\omega) \in \mathbb{R}^{n \times n} \tag{4.11}$$

and

$$S = \begin{bmatrix} 1 & 1 & \cdots & 1 \\ 1 & 1 & \cdots & 1 \\ \vdots & & & \\ 1 & 1 & \cdots & 1 \end{bmatrix} \in \mathbb{R}^{(m+1) \times n} \tag{4.12}$$

It can be seen from Eqs. (4.9), (4.10) that the invariant matrix of the reparametrized system is

$$\tilde{\mathcal{M}} = \tilde{B} \cdot \tilde{A} = (\bar{B} + S \cdot \Omega) \cdot \tilde{A} \tag{4.13}$$

Therefore, the matrix inequality for examining the global stability of the reparametrized system is the following

$$-C < 0 \tag{4.14}$$
$$\tilde{\mathcal{M}}^T \cdot C + C \cdot \tilde{\mathcal{M}} \leq 0 \tag{4.15}$$

that is,

$$-C < 0 \tag{4.16}$$
$$\tilde{A}^T(\bar{B}^T + \Omega S^T)C + C(\bar{B} + S\Omega)\tilde{A} \leq 0 \tag{4.17}$$

which clearly has the same form as Eq. (4.6), but with the following set of unknowns:

$$
x = \begin{bmatrix} x_1 \\ x_2 \\ \vdots \\ x_{m+1} \\ x_{m+2} \\ \vdots \\ x_{m+n+1} \end{bmatrix} = \begin{bmatrix} c_1 \\ c_2 \\ \vdots \\ c_{m+1} \\ \omega_1 \\ \vdots \\ \omega_n \end{bmatrix}
\tag{4.18}
$$

that makes it a bilinear matrix inequality. Now we are ready to construct the parameter matrices in the BMI in Eq. (B.19) starting with

$$
G_0^1 = G_0^2 = 0^{(m+1)\times(m+1)}
\tag{4.19}
$$

$$
G_{ki,j}^1 = \begin{cases} -1, & i = j = k \\ 0, & \text{otherwise} \end{cases}
\tag{4.20}
$$

$$
i, j, k = 1, \ldots, m+1
$$

$$
G_k^1 = 0^{(m+1)\times(m+1)}, \quad k = m+2, \ldots, m+n+1
\tag{4.21}
$$

and

$$
K_{kl}^1 = 0^{(m+1)\times(m+1)}, \quad k, l = 1, \ldots, m+n+1
\tag{4.22}
$$

Furthermore, let us introduce the following notations

$$
P_k \in \mathbb{R}^{(m+1)\times(m+1)}
$$

$$
P_{ki,j} = \begin{cases} \bar{B} \cdot \tilde{A}_{i,j}, & i = k \\ 0, & i \neq k \end{cases} \quad i, j, k = 1, \ldots, m+1
\tag{4.23}
$$

and

$$
Q_{kl} \in \mathbb{R}^{(m+1)\times(m+1)}
$$

$$
Q_{kli,j} = \begin{cases} \tilde{A}_{l-m-1,j}, & i = k \\ 0, & i \neq k \end{cases}
\tag{4.24}
$$

$$
i, j, k = 1, \ldots, m+1, \ l = m+2, \ldots, m+n+1
$$

Then

$$
G_k^2 = \begin{cases} P_k + P_k^T, & k = 1, \ldots, m+1 \\ 0^{(m+1)\times(m+1)}, & k = m+2, \ldots, m+n+1 \end{cases}
\tag{4.25}
$$

and

$$K_{kl} = \begin{cases} Q_{kl} + Q_{kl}^T, & k = 1, \ldots, m+1, \, l = m+2, \ldots, m+n+1 \\ 0^{(m+1) \times (m+1)}, & \text{otherwise} \end{cases}$$

$$(4.26)$$

$$k, l = 1, \ldots, m+n+1$$

We note that in certain cases the feasibility of a BMI can be traced back to the feasibility of equivalent LMIs (see [38] or [39]), but in our case it is not possible because of the structural (diagonality) constraint on both of the unknown matrices Ω and C in Eq. (4.17).

4.1.3.2 Example
In order to illustrate the previous-proposed method of finding time-reparametrization transformations for global stability analysis, a simple numerical example is presented here.

Consider a QP system with the following matrices

$$\mathcal{A} = \begin{bmatrix} \frac{2}{3} & -\frac{8}{3} \\ \frac{2}{3} & -\frac{7}{3} \end{bmatrix} \approx \begin{bmatrix} 0.6667 & -2.6667 \\ 0.6667 & -2.3333 \end{bmatrix} \qquad (4.27)$$

$$\mathcal{B} = \begin{bmatrix} \frac{2}{3} & -\frac{1}{3} \\ -\frac{8}{3} & \frac{16}{3} \end{bmatrix} \approx \begin{bmatrix} 0.6667 & -0.3333 \\ -2.6667 & 5.3333 \end{bmatrix} \qquad (4.28)$$

$$\lambda = \begin{bmatrix} 2 \\ \frac{5}{3} \end{bmatrix} \approx \begin{bmatrix} 2 \\ 1.6667 \end{bmatrix} \qquad (4.29)$$

Its equilibrium point of interest is:

$$x^* = [1 \ 1]^T \qquad (4.30)$$

The Jacobian matrix of the locally linearized system in x^* has the following eigenvalues: -0.1187, -4.9924. This shows that the investigated equilibrium point is at least locally asymptotically stable.

Using an appropriate LMI solver (e.g., Matlab's LMI Control Toolbox) it can be checked that the LMI (Eq. 4.6) cannot be solved for $\mathcal{M} = \mathcal{B} \cdot \mathcal{A}$. However, using the algorithm [40] for solving the corresponding BMI we find that a feasible solution of Eq. (4.17) is, for example,

$$C = \begin{bmatrix} 1 & 0 & 0 \\ 0 & 1 & 0 \\ 0 & 0 & 1 \end{bmatrix}, \quad \omega = \begin{bmatrix} \frac{2}{3} & -\frac{5}{3} \end{bmatrix}^T \tag{4.31}$$

The eigenvalues of $\tilde{M}^T \cdot C + C \cdot \tilde{M}$ are

$$\lambda_1 = 0, \quad \lambda_2 \approx -0.2374, \quad \lambda_3 \approx -9.9848 \tag{4.32}$$

which shows that the examined system is globally stable.

4.2 STABILITY OF KINETIC SYSTEMS

The model form and basic properties of kinetic systems were introduced in Section 2.3, where it was shown that the structure of a kinetic system (with not necessarily mass action kinetics) is characterized by its reaction graph and by its deficiency. Both the *reaction graph and the deficiency are parameter independent*, that is, they do not depend on the actual values of the positive kinetic coefficients k_{ij}.

In Section 2.3.2.2, a special entropy-line Lyapunov function was also given in the form

$$V(x) = \sum_{i=1}^{n} x_i \left(\ln\left(\frac{x_i}{x_i^*} \right) - 1 \right) + x_i^* \tag{4.33}$$

where x^* denotes the equilibrium point in the given stoichiometric compatibility class, that is also *kinetic parameter independent*.

4.2.1 Deficiency Zero and Deficiency One Theorems

The celebrated Deficiency Zero and Deficiency One theorems give conditions related to stability (see Section B.4.2 on stability notions and analysis) of kinetic systems.

The exact forms of the Deficiency Zero and Deficiency One theorems are taken from [16], where the term "reaction network" is used instead of kinetic system.

Theorem 1 (Deficiency zero theorem). *For any reaction network of deficiency zero the following statements hold true:*

1. *If the network is not weakly reversible then, for arbitrary kinetics (not necessarily mass action), the differential equations for the corresponding*

reaction system cannot admit a positive steady state (i.e., a steady state in $\mathbb{R}_+^n$).

2. *If the network is not weakly reversible then, for arbitrary kinetics (not necessarily mass action), the differential equations of the corresponding reaction system cannot admit a cyclic composition trajectory along which all species concentrations are positive.*

3. *If the network is weakly reversible then, for mass action kinetics (but regardless of the positive values the reaction rate coefficients take), the differential equations of the corresponding reaction system have the following properties: There exists within each positive stoichiometric compatibility class precisely one steady state; that steady state is asymptotically stable; and there is no nontrivial cyclic composition trajectory along which all species concentrations are positive.*

Theorem 2 (Deficiency one theorem). *Consider a mass action system for which the underlying reaction network has ℓ linkage classes, each containing just one terminal strong linkage class. Suppose that the deficiency d of the network and the deficiencies of the individual linkage classes d_i, $i = 1, \ldots, \ell$ satisfy the following conditions:*

1. $d_i \leq 1, \quad i = 1, \ldots, \ell$
2. $\sum_{i=1}^{\ell} d_i = d$

Then, no matter what (positive) values the reaction rate coefficients take, the corresponding differential equations can admit no more than one steady state within a positive stoichiometric compatibility class. If the network is weakly reversible, the differential equations admit precisely one steady state in each positive stoichiometric compatibility class.

The earlier theorems establish very *strong results about the qualitative dynamical properties of kinetic systems* that are also *robust with respect to the parameters (i.e., the reaction rate coefficients) and depend on the structure and on the stoichiometry of the system* only.

Remark

1. The weakly reversible kinetic systems in the Deficiency Zero Theorem admit a logarithmic Lyapunov function in the form of Eq. (4.33).
2. We add the important fact that the stability of the system in the case of the Deficiency Zero Theorem is global if the CRN has one linkage class [41], since the deficiency zero property and weak reversibility implies complex balance (see also the following section).

3. It is also worth mentioning the following significant control-related results on weakly reversible deficiency zero networks. In [42] the detectability of such systems was studied and nonlinear observers were proposed for them with proven convergence under mild conditions. It was shown in [43] that weakly reversible deficiency zero networks are input-to-state stable if the (nonvanishing) manipulable inputs are the reaction rate coefficients.
4. It must be stressed that by distancing ourselves from the original chemical motivations, we consider kinetic models as a general nonlinear system class, and do not require in general that such models describe a chemically strictly feasible reaction mechanism. Thus, it is allowed that certain thermodynamical constraints (such as component mass balance conservation) are not fulfilled in the examined models.
5. We have seen in Section 2.3.3, that a given kinetic ODE may admit more than one structurally different dynamically equivalent realizations with different deficiency and reversibility properties. Therefore, it is of great importance even from stability analysis viewpoint to be able to search for realizations with given properties, that will be the subject of Section 4.5.

Example 11 (Stability of the mass action model of the simple reversible Michaelis-Menten kinetics). In order to illustrate the use of the Deficiency Zero and Deficiency One theorems earlier, let us consider the reversible version of the simple Michaelis-Menten kinetics obeying the mass action law, that has been introduced in Example 5.

The reaction graph of the model is presented in Fig. 4.1 with the reaction rate coefficients $K_i^{+,-}$ as edge weights.

This defines the complex composition matrix Y and the corresponding monomial vector ψ as

$$
Y = \begin{bmatrix} 1 & 0 & 0 \\ 1 & 0 & 1 \\ 0 & 1 & 0 \\ 0 & 0 & 1 \end{bmatrix}, \quad \psi(x) = \begin{bmatrix} x_1 x_2 \\ x_3 \\ x_2 x_4 \end{bmatrix}
$$

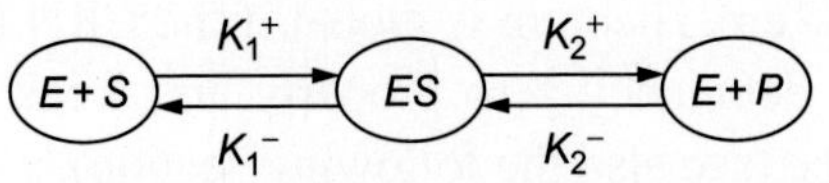

Fig. 4.1 The reaction graph of the mass action-type model of the simple reversible Michaelis-Menten kinetics.

It is seen in Fig. 4.1 that the reaction graph is reversible, thus the system admits a positive equilibrium point in the state space. We have $m = 3$ complexes, $l = 1$ linkage classes, and the dimension of the stoichiometric subspace s is 2; therefore, the deficiency of this realization is $\delta = m-l-s = 3 - 2 - 1 = 0$.

Therefore, the Deficiency Zero theorem applies in this case, and we conclude that *this model has within each positive stoichiometric compatibility class precisely one steady state that is asymptotically stable.*

4.3 INVARIANTS (FIRST INTEGRALS) FOR QP AND KINETIC SYSTEMS

A function $I: \mathbb{R}^n \mapsto \mathbb{R}$ of an autonomous ordinary differential equation $\frac{dx(t)}{dt} = f(x)$ with $x(t) \in \mathbb{R}^n$ is called and *invariant or a first integral* of this equation if

$$\frac{d}{dt}I = \frac{\partial I}{\partial x} \cdot \frac{dx(t)}{dt} = 0 \qquad (4.34)$$

This means that the value of I is constant along the trajectory of system evolution (i.e., it corresponds to a conserved quantity of some kind). Moreover, the manifold determined by the equation $I(x(t)) = const$ represents the manifold in the state space over which the system state moves.

4.3.1 Linear First Integrals of Kinetic Systems and Their Relations to Conservation

The notion and basic properties of kinetic systems were introduced in Section 2.3. Their dynamic model was based upon the so-called elementary reactions (see Eq. 2.19) that is used here in the following form

$$\sum_{i=1}^{n} \alpha_{ki}X_i \rightarrow \sum_{i=1}^{n} \alpha_{ji}X_i \qquad (4.35)$$

where X_i, $i = 1, \ldots, n$ denotes a specie, and the positive integers α_{ki} are the so-called stoichiometric coefficients. The complex composition matrix Y is formed from the stoichiometric coefficients such that $Y_{ij} = \alpha_{ji}$, $i = 1, \ldots, n$, $j = 1, \ldots, m$.

Assuming mass action kinetics, the dynamic model of the kinetic system is in the form

$$\dot{x} = M \cdot \psi(x) \tag{4.36}$$

where $x \in \mathbb{R}^n$ is the state variable and $M \in \mathbb{R}^{n \times m}$. The monomial vector function $\psi: \mathbb{R}^n \to \mathbb{R}^m$ is defined as

$$\psi_j(x) = \prod_{i=1}^{n} x_i^{Y_{ij}}, \quad j = 1, \ldots, m \tag{4.37}$$

A pair (Y, A_k) is called a realization of Eq. (4.36), if $M = YA_k$. One can define the stoichiometric matrix $\mathbf{S} \in \mathbb{Z}^{n \times r}$ of a realization (Y, A_k) where r is the number of the reactions and $\mathbf{S}$ has a column $Y_{\cdot j} - Y_{\cdot i}$ iff $A_{k\,ij} > 0$ for all $i, j = 1, \ldots, m$.

4.3.1.1 Mass Conserving Chemical Reactions

Let us define g_v as the molecular weight of the specie X_v with strictly positive value. If reaction $C_i \to C_j$ is present in the network, the following can be written:

$$\sum_{v=1}^{n} \alpha_{vi} g_v = \sum_{v=1}^{n} \alpha_{vj} g_v = c_s \tag{4.38}$$

where $c_s > 0$ is a constant column-sum and $\alpha_{vi} = Y_{vi}$, $\alpha_{vj} = Y_{vj}$ according to the definition. Let us define vector $g \in \mathbb{R}_+^n$ as a row vector formulated from the molecular weights. Now Eq. (4.38) can be rewritten as $g \cdot Y_{\cdot i} = g \cdot Y_{\cdot j} = c_s$ where $Y_{\cdot i}$ refers to the ith column of matrix Y, or with other words, the composition vector of complex C_i. Finally, it can be said that *a reaction is mass conservative* if the following holds:

$$g \cdot \eta^{(i,j)} = 0 \tag{4.39}$$

where $\eta^{(i,j)} = Y_{\cdot i} - Y_{\cdot j}$ is the *reaction vector* and g is strictly positive.

In [44] the *mass conservative reaction set* is defined as a set of the reactions having the earlier property. It should be noted that a *given CRN is called mass conservative if all of its reactions are in the mass conservative reaction set and a common g can be determined.*

In [45] the notion of *stoichiometrically mass conserving* kinetic systems were introduced that have a positive vector in the left kernel of their stoichiometric matrix.

$$\mathbf{S}^T a = 0 \tag{4.40}$$

where $a \in \mathbb{R}^n$, $a_i > 0$, $i = 1, \ldots, n$, and $\mathbf{S}$ is the stoichiometric matrix.

Because of the possible existence of dynamically equivalent realizations of kinetic systems with different reaction sets; however, *the property of mass conservation in the sense of Eq. (4.39) is realization dependent.*

Example 12 (Linear kinetic systems). Since linear kinetic systems have $Y = I$ and $M = A_k$ with a *unique realization*, they are all mass conserving with $a = [1, 1, \ldots, 1]^T$. This means that the molecular weights of all complexes—or equivalently all species—are the same.

4.3.1.2 Linear First Integrals

An alternative *realization-independent but formal definition of mass conservation* builds on the notion of fist integrals. The *kinetic system* (4.36) *is called mass conserving if and only if it has a linear first integral with positive coefficients* in the form

$$I(x) = a^T x \tag{4.41}$$

where $a \in \mathbb{R}^n$ and $a_i > 0$, $i = 1, \ldots, n$.

From the previous definition it follows immediately that *mass conserving kinetic systems have bounded trajectories.*

It is easy to see that a sufficient condition of mass conservation in the sense of Eq. (4.41) can be given as follows. *If there exists a positive vector in the left kernel of M then the kinetic system is mass conserving,* because

$$\nabla(a^T x)M\psi(x) = \underbrace{a^T M}_{0}\psi(x) = 0 \tag{4.42}$$

where $a \in \mathbb{R}^n$ and $a_i > 0$, $i = 1, \ldots, n$.

4.3.2 Invariants of QP Systems and Their Computation

The invariants or first integrals of QP systems can be retrieved by using the logarithmic form of QP models (see Eq. 3.38) in their canonical LV form in Eq. (3.41)

$$\frac{d \ln x}{dt} = \Lambda + \mathcal{M}x \tag{4.43}$$

where the state variable vector x is now the vector of quasimonomial p of the original equation.

By using variable embedding and QM transformation described in Section 3.3, $m - n$ nonlinear algebraic equations can be derived between the state variables of Eq. (4.43) in the $m > n$ case (where m is the number of quasimonomials and n is that of the state variables of the original QP model)

$$x_i^+ = \prod_{j=1}^{n} (x_j^*)^{\mathcal{B}_{ij}^{**}}, \quad i = 1, \ldots, m - n \tag{4.44}$$

where $x^* \in \mathbb{R}^n$ is a partition of the state vector such that the corresponding partition of the full rank exponent matrix rank $\mathcal{B} = n$ of the original QP model $\mathcal{B}^* \in \mathbb{R}^{n \times n}$ is an invertible matrix, and the remaining partitions x^+ and $\mathcal{B}^+$ satisfy

$$\mathcal{B}^{**} = \mathcal{B}^+ (\mathcal{B}^*)^{-1}$$

From the previous equations and from the definition in Eq. (4.34), it is clear that Eq. (4.44) defines $(m - n)$ quasipolynomial invariants of the LV model (4.43) in the form

$$I_i(x) = x_i^+ - \prod_{j=1}^{n} (x_j^*)^{\mathcal{B}_{ij}^{**}}, \quad i = 1, \ldots, m - n \tag{4.45}$$

Based on the previous notions and results, a numerically effective polynomial-time algorithm is proposed in the paper [46] for the determination of a class of first integrals in quasipolynomial systems. The algorithm was implemented and tested in the Matlab numeric computational software environment.

4.4 RELATIONS BETWEEN THE LYAPUNOV FUNCTIONS OF QP AND KINETIC MODELS

In Sections 4.1 and 4.2 the results related to the stability analysis of QP and kinetic systems were discussed, and the special Lyapunov functions of these positive polynomial systems were introduced. It is apparent that the Lyapunov function candidate of QP systems in Eq. (4.4) and that of kinetic systems in Eq. (4.33) are similar in their form, but they are not the same. This section summarizes the results available about the relationships between these Lyapunov functions based mainly on [47].

4.4.1 The Physical-Chemical Origin of the Natural Lyapunov Function of Kinetic Models

Kinetic systems originate from a special class of process systems, where chemical reactions take place under isothermal and isobaric conditions in a perfectly stirred tank reactor [48]. This fact can be applied to consider the entropy $\mathbb{S}$ of this reactor as a natural Lyapunov function candidate for stability analysis, if the dynamic system model originates from conservation balances.

It is easy to see that the dynamic model of kinetic systems in Eqs. (4.36), (4.37) originates from dynamic component mass balances for the species X_i, $i = 1, \ldots, n$ expressed as differential equations for species concentrations as state variables x_i; therefore, the entropy of the system as a function of the state variables can be written as [49]

$$\mathbb{S}(x) = -K_S \sum_{i=1}^{n} x_i \ln\left(\frac{x_i}{x_i^*}\right) \tag{4.46}$$

where x_i^* is the steady-state value of the state variable x_i, and K_S is a positive constant. It is important to remark that $\mathbb{S}$ *does not depend directly on the system parameters (neither on the coefficient matrix M nor on the complex composition matrix Y), that is, it is realization independent.* Moreover, $\mathbb{S}$ is additive for the species' concentrations, that is

$$\mathbb{S}(x) = \sum_{i=1}^{n} \mathbb{S}(x_i)$$

with $\mathbb{S}(x_i) = -K_S x_i \ln\left(\frac{x_i}{x_i^*}\right)$.

In order to comply with the necessary properties of a Lyapunov function, a slightly modified version of the entropy (4.46) is used for kinetic systems in the form

$$V_{\text{CRN}}(x) = \sum_{i=1}^{n} x_i \left(\ln x_i - \ln x_i^*\right) + \left(x_i^* - x_i\right) \tag{4.47}$$

that is equivalent with Eq. (4.33). Similar to the originating entropy function $\mathbb{S}$, *the Lyapunov function candidate V_{CRN} does not depend on the system parameters, and it is additive for the state variables.*

4.4.2 Relationship With the Logarithmic Lyapunov Function of QP Models

We have already seen in Section 3.3 that LV systems are representative elements of QP model equivalence classes that are invariant under QM transformation. Their dynamic model in its logarithmic form (see Eq. 3.41) contains the invariant parameters $\mathcal{M}$ and Λ

$$\frac{d\ln x}{dt} = \Lambda + \mathcal{M}x \tag{4.48}$$

where $x \in \mathbb{R}^m$ is the state variable of the LV model, that is the same as the vector of quasimonomials p of the original QP models belonging to the equivalence class.

The candidate Lyapunov function family [13, 14] is the same for the whole QP model equivalence class, and it is expressed as a function of the LV model's state variables

$$V_{\mathrm{LV}}(x) = \sum_{i=1}^{m} c_i \left[-x_i^* \left(\ln x_i - \ln x_i^* \right) + \left(x_i - x_i^* \right) \right] \tag{4.49}$$

where $c_i > 0$, $i = 1, \ldots, m$ are positive constants, and x_i^* is the steady-state value of the state variable x_i. It is important to note that *the Lyapunov function candidate V_{LV} does not depend on the system parameters, and it is also additive for the state variables.*

Comparing the candidate Lyapunov functions (4.47), (4.49), we can observe that they have a very similar algebraic form. Other than a constant rescaling, they only differ in the multiplicative factor x_i versus x_i^* in the first term.

In the pioneering paper of Gorban et al. [50], the level set equivalence of various entropy functions (including the ones in the form of V_{CRN} and V_{LV} earlier) was proved for Markov chains with positive equilibrium point. This, and the mathematical equivalence of Markov chains and linear weakly reversible kinetic systems with mass action kinetics, was applied in [47] to show that linear kinetic systems with a linear first integral admit both the traditional Eq. (4.47) and the LV type (4.49) Lyapunov function candidate that are level set equivalent.

Furthermore, we have seen in Section 2.4 that LV models can always be represented in a kinetic form. This opens up further possibilities to prove the level set equivalence for wider QP or CRN model subclasses.

4.5 COMPUTATIONAL ANALYSIS OF THE STRUCTURE OF KINETIC SYSTEMS

As we have seen in Section 2.3 on kinetic systems obeying mass action law, their dynamic model is a set of positive polynomial ODEs characterized by the monomial vector $\psi(x)$ and the coefficient matrix M of the form of Eq. (2.46):

$$\dot{x} = M \cdot \psi(x) \tag{4.50}$$

In Section 2.3.3, the notion of dynamical equivalence and the realization of mass action law kinetic systems were introduced, and it was shown that a given kinetic system model may admit different realizations, that is, different reaction graphs (or equivalently different A_k matrices).

In this section, we briefly present an optimization-based framework for computing reaction graphs (realizations) based on the ODE description of kinetic models.

4.5.1 Computational Model for Determining Linearly Conjugate Realizations of Kinetic Systems

The concept of linear conjugacy has been already introduced in Section 3.1.3 in relation with the constant positive diagonal transformation of CRNs with a diagonal matrix D. It is easy to see that dynamical equivalence is a special case of linear conjugacy with $D = I$ (with I being a unit matrix), so we describe the more general linear conjugacy case here. Linearly conjugate realizations can be computed by using a linear optimization model [31]. As it was described in Section 3.1, Eq. (3.12) in the form of $Y \cdot A_k = D \cdot M'$ must be fulfilled to ensure linear conjugacy. Moreover, the Kirchhoff property of A_k and the positivity of the transformation parameters in D have to be ensured.

These constraints can be expressed by the following equations:

$$D \cdot M' - Y \cdot A_k = \mathbf{0} \tag{4.51}$$

$$\underline{1} \cdot A_k = 0 \tag{4.52}$$

$$A_{k\,ij} \geq 0, \quad i, j \in \{1, \ldots, m\}, \ i \neq j \tag{4.53}$$

$$D_{ii} > 0, \quad i \in \{1, \ldots, n\} \tag{4.54}$$

where $\mathbf{0} \in \mathbb{R}^{n \times m}$ denotes the zero matrix, and $\underline{1} \in \mathbb{R}^m$ is a vector, each entry of which is one. The matrices Y and M' are known, and the *decision variables* are contained in the matrices D and A_k. Specifically, the variables are the off-diagonal entries of the matrix A_k and the diagonal entries of the matrix D. The diagonal entries of A_k are determined by the column conservation property described in Eq. (4.52).

To compute a CRN structure with a desired property, often additional linear constraints are also needed in the form of linear equalities and inequalities. These will generally be denoted as

$$L(A_k, D) \leq 0 \tag{4.55}$$

Whenever it is possible, we will simply denote the set of the above linear (in)equalities by L.

One frequently used example for L applied in the presented computational methods that it might be necessary to exclude some set $\mathcal{H} \subset \mathcal{R}$ of reactions from the computed realizations. This can be written in the form of a linear constraint as follows:

$$A_{kji} = 0 \quad (C_i, C_j) \in \mathcal{H} \tag{4.56}$$

The feasibility of the linear constraints (4.51)–(4.55) can be verified, and valid solutions (if they exist) can be determined in the framework of linear programming.

It is important to remark that in the earlier computation model, the boundedness property of all variables can be ensured so that the set of possible reaction graphs remains the same as it was proved in [51]. In other words, we do not lose any valid solution if we set upper bounds for the off-diagonal elements of A_k.

A more detailed description about the basics of the applied computational tools can be found in Section B.3.

4.5.2 Dense and Sparse Realizations and Their Properties
Here we formally define the notion of dense and sparse realizations and describe their most important properties. The results summarized in this section are based on [24, 51–55].

Assuming a fixed complex set given by the matrix Y, a realization of a kinetic system is a *dense (sparse) realization* if the maximum (minimum) number of reactions take place. It is easy to see that the dense (sparse) property of a CRN realization is equivalent to the feature that its Kirchhoff matrix A_k has the maximum (minimum) number of positive off-diagonal entries.

4.5.2.1 Dense Reaction Graphs
In a set of directed graphs we call a graph a *superstructure* if it contains each element as a subgraph (not considering the edge weights) and it is minimal under inclusion. It is clear that the superstructure is unique, because there cannot be two different graphs that are subgraphs of each other. It was proved in [53] that the dense realizations determine superstructures among dynamically equivalent and linearly conjugate realizations. This means

that the reaction graphs of all the possible dynamically equivalent/linearly conjugate realizations on the same vertex set—not considering the weights of the edges—are subgraphs of the reaction graph of the corresponding dense realization. It is easy to extend this result to the case when additional linear constraints are given in the form of Eq. (4.55). Therefore, the superstructure property of dense realizations can be summarized by the following proposition taken from [51].

Proposition 1. *Among all the realizations linearly conjugate to a given kinetic system of the form (2.46) on a fixed set of complexes and fulfilling a finite set of additional linear constraints, the dense realization with the prescribed properties determines a superstructure.*

The *dense linearly conjugate realization of a kinetic system can be determined in polynomial time* as follows. All the possible solutions of an LP problem (see Section B.3.1) are points of a convex (closed) polyhedron $\mathcal{P}$, which is the intersection of the closed half-planes determined by the constraints. In a standard LP problem, the half-planes are closed because all the constraints are nonstrict inequalities. Our original computation model contains strict inequalities (see Eq. 4.54), but these can be transformed to nonstrict ones without losing any solution [51].

According to the number of variables the polyhedron $\mathcal{P}$ is in $\mathbb{R}^{m^2-m+n}$. Let the point describing the linearly conjugate realization (D^i, A_k^i) be $P^i = (p_1^i, \ldots, p_n^i, \ldots, p_{m^2-m+n}^i) \in \mathcal{P}$ so that the first n coordinates represent the diagonal entries of matrix D and the rest are the off-diagonal entries of matrix A_k according to columns.

In the *algorithm that determines dense realizations*, we use the following subroutine:

FindPositive(M, Y, L, H) returns a point $Q \in \mathcal{P}$ that fulfils the model determined by matrices M and Y and a finite set L of nonstrict linear inequalities, so that considering a set H of indices, the value of the objective function $\sum_{j \in H} q_j$ is maximal, and $Q = [q_1 \ \ldots \ q_{m^2-m+n}]^T$.

This procedure also returns the set B of indices where $k \in B$ if and only if $q_k > 0$.

The computation can be performed in polynomial time because it requires the solution of an LP problem and the checking of the elements in a set of size $m^2 - m + n$.

Now we can give the pseudocode of the algorithm for computing dense realizations. The correctness proof of the method can be found in [51].

Algorithm 1   Computes the dense linearly conjugate realization of a kinetic system

Inputs: M, Y, L
Output: *Result*

1: $H := \{1, 2, \ldots, m^2 - m + n\}$
2: $B := H$
3: $Result := \mathbf{0} \in \mathbb{R}^{m^2 - m + n}$
4: $loops := 0$
5: **while** $B \neq \emptyset$ **do**
6:      $(Q, B) := \text{FindPositive}(Y, M, L, H)$
7:      $Result := Result + Q$
8:      $H := H \setminus B$
9:      $loops := loops + 1$
10: **end while**
11: $Result := Result/loops$
12: **if** $\exists i \in \{1, \ldots, n\} \cap H$ **then**
13:      There is no linearly conjugate realization of the kinetic system
14:      (M, Y) fulfilling the set L of constraints.
15: **else**
16:      *Result* determines a dense linearly conjugate realization of the
17:      kinetic system (M, Y) fulfilling the L of constraints.
18: **end if**

4.5.2.2 Sparse Realizations

For the exact computation of sparse structures, we have to track individually whether an off-diagonal element of the Kirchhoff matrix is practically zero (i.e., it is below the zero threshold denoted by ϵ) or not. For this, we introduce the binary variables $\delta_{ij} \in \{0, 1\}$ for $i, j = 1, \ldots, m$, $i \neq j$. Moreover, for practical reasons, we introduce upper bounds u_{ij} for the elements of the Kirchhoff matrix, that is

$$A_{k\,ij} \leq u_{ij}, \quad i, j = 1, \ldots, m, \quad i \neq j \tag{4.57}$$

We would like to prescribe that $\delta_{ij} = 1$ if and only if $A_{k\,ij} \geq \epsilon$. We will denote this condition as follows: $\delta_{ij} = 1 \leftrightarrow A_{k\,ij} \geq \epsilon$. One can easily check that the following inequalities are equivalent to this condition; that is, they are fulfilled if and only if either $A_{k\,ij} \geq \epsilon$ and $\delta_{ij} = 1$, or $A_{k\,ij} < \epsilon$ and $\delta_{ij} = 0$:

$$A_{k\,ij} - \epsilon\delta_{ij} \geq 0, \quad i,j = 1,\ldots,m, \quad i \neq j \tag{4.58}$$

$$u_{ij}\delta_{ij} - A_{k\,ij} \geq 0, \quad i,j = 1,\ldots,m, \quad i \neq j \tag{4.59}$$

Therefore, a sparse linearly conjugate realization of a kinetic system can be computed by using the constraints (4.51)–(4.54), (4.58), and (4.59) and minimizing the objective function

$$F_{sp} = \sum_{\substack{i,j \\ i \neq j}} \delta_{ij} \tag{4.60}$$

which is a standard mixed integer linear programming (MILP) problem.

We remark that for many kinetic systems (especially large ones) fulfilling certain technical conditions, sparse realizations can be computed in polynomial time without integer variables [56].

4.5.2.3 Analyzing Structural Uniqueness of CRN Structures Using Dense and Sparse Realizations

Network inference (i.e., the computation of network structure/parameters using measurement data and prior information) is a common task in systems biology. The parameters of kinetic models are often estimated from concentration data in the form of a polynomial equation (4.50), since the right-hand side is a linear function of M. In such cases, it might be an important question whether the structure of the reaction network corresponding to the identified dynamical model is unique or not. Luckily, we can give an answer to this question using our computational framework as follows.

Proposition 2. *The dense and sparse realizations of a kinetic system [M, Y] have the same number of reactions if and only if all realizations of the model are structurally identical.*

4.5.3 Computing Linearly Conjugate Realizations With Preferred Properties

We have already seen in Section 2.3.2.2 that the most important structural properties determining the qualitative dynamic behavior of a kinetic system, namely the deficiency and the reversibility are realization-dependent, so it is of primary importance from the viewpoint of model analysis and controller design to find dynamically equivalent or linearly conjugate realizations with prescribed (preferred) structural properties.

4.5.3.1 Weakly Reversible Structures

It is known that the weak reversibility of the reaction graph is equivalent to the property that A_k contains a positive vector in its kernel. This can be written as

$$A_k \cdot p = 0 \qquad (4.61)$$

where p is an element-wise strictly positive vector. Since p is unknown, condition (4.61) is not a linear constraint. Therefore, let us introduce a scaled Kirchhoff matrix

$$\bar{A}_k = A_k \cdot \mathrm{diag}(p) \qquad (4.62)$$

Clearly, condition (4.61) can be written as $A_k \cdot \mathrm{diag}(p) \cdot \underline{1} = 0$ which gives

$$\bar{A}_k \cdot \underline{1} = 0 \qquad (4.63)$$

It is easy to see that the structure (i.e., the positions of the zero and nonzero elements) of A_k and $\bar{A}_k$ is the same. This can be simply ensured by introducing binary variables δ_{ij} for $i,j = 1,\ldots,m$, $i \neq j$ and prescribing the relations

$$\delta_{ij} = 1 \leftrightarrow A_{k\,ij} > \epsilon, \quad i,j = 1,\ldots,m, \quad i \neq j \qquad (4.64)$$
$$\delta_{ij} = 1 \leftrightarrow \bar{A}_{k\,ij} > \epsilon, \quad i,j = 1,\ldots,m, \quad i \neq j \qquad (4.65)$$

Similar to Eqs. (4.58), (4.59), condition (4.65) can be translated into the following linear inequalities

$$\bar{A}_{k\,ij} - \epsilon \delta_{ij} \geq 0, \quad i,j = 1,\ldots,m, \quad i \neq j \qquad (4.66)$$
$$\bar{u}_{ij}\delta_{ij} - \bar{A}_{k\,ij} \geq 0, \quad i,j = 1,\ldots,m, \quad i \neq j \qquad (4.67)$$

where $\bar{u}_{ij}$ is the positive upper bound for $[\bar{A}_k]_{ij}$. In summary, the constraints for computing linearly conjugate weakly reversible realizations (if they exist) are the following: Eqs. (4.51)–(4.54), (4.58), (4.59), (4.63), (4.66), and (4.67). Then, a dense or sparse weakly reversible realization can be computed by maximizing or minimizing the objective function F_{sp} given in Eq. (4.60), respectively. This can be performed again in the framework of MILP.

It is important to add that the computation of weakly reversible structures can alternatively be performed in several LP steps without integer variables, using elementary graph theory [51].

4.5.3.2 Complex Balanced Realizations

Let us recall first the condition of complex balance in Eq. (2.44) which is clearly a linear constraint if any equilibrium point x^* is known. The equilibrium point of the linearly conjugate CRN is given by

$$x^{'*} = D \cdot x^* \tag{4.68}$$

The complex balance condition for the linearly conjugate network is

$$A'_k \cdot \psi(x^{'*}) = 0 \tag{4.69}$$

Using the notations of Section 3.1.3, the left-hand side of Eq. (4.69) can be further written as

$$A'_k \cdot \psi(x^{'*}) = A'_k \cdot \psi(D \cdot x^*) = A_k \cdot \Phi_{D^{-1}} \cdot \Phi_D \cdot \psi(x^*) \tag{4.70}$$

Therefore, the linear condition for the complex balance property in the linearly conjugate case is simply

$$A_k \cdot \psi(x^*) = 0 \tag{4.71}$$

and the whole constraint set for the computation is Eqs. (4.51)–(4.54) and (4.71).

4.5.3.3 Deficiency Zero Realizations

Let us consider a kinetic polynomial system of the form (2.46) and a fixed complex set defined by matrix Y. As it is shown in [54], it is possible to compute using optimization a deficiency zero realization for the system, if it exists. We briefly summarize these results below.

The general case: Here the only goal is to compute a deficiency zero reaction graph without any other requirements. The linear constraints for this in addition to Eqs. (4.51)–(4.54) are given by

$$\tilde{y}^{(\ell)} = \eta^{(\ell)} + Y^T \cdot \alpha^{(\ell)}, \quad \ell = 1, \ldots, m - \text{rank}(Y) \tag{4.72}$$

$$A_{k\,ij} \leq U_1 \cdot \Theta_{ij} \quad i, j = 1, \ldots, m \tag{4.73}$$

$$|\eta_i^{(\ell)} - \eta_j^{(\ell)}| \leq 2 \cdot U_2(1 - \Theta_{ij}) \quad i, j = 1, \ldots, m, \quad \ell = 1, \ldots, m - \text{rank}(Y) \tag{4.74}$$

The new continuous decision variables in Eqs. (4.72)–(4.74) are $\alpha^{(\ell)}$ and $\eta_i^{(\ell)}$ for $\ell = 1, \ldots, m - \text{rank}(Y)$ and $i = 1, \ldots, m$. Moreover, Θ_{ij} for

$i, j = 1, \ldots, m$ are new binary decision variables. The additional constant parameters in Eqs. (4.72)–(4.74) are the following: $\tilde{y}^{(\ell)}$ are arbitrary kernel basis vectors of Y, while U_1 and U_2 are upper bounds for the absolute values of the entries of A_k and $\eta^{(\ell)}$, respectively. It is clear form the previous constraints that deficiency zero realizations can be computed using an MILP solver.

The weakly reversible case: The aim here is to compute deficiency zero weakly reversible realizations. If such a realization exists, the stability implications of the Deficiency Zero Theorem can be applied for the dynamical system (see Section 2.3.2.2). A necessary condition for the existence of a weakly reversible linearly conjugate realization is

$$M \cdot p = 0, \quad p \in \mathbb{R}^n_+ \tag{4.75}$$

Clearly, Eq. (4.75) can be checked via LP. If Eq. (4.75) is fulfilled, it can be shown that the only constraint that has to be added to Eqs. (4.51)–(4.54) to compute weakly reversible deficiency zero realizations is

$$A_k \cdot \eta^{(i)} = 0, \quad i = 1, \ldots, m - \text{rank}(M) \tag{4.76}$$

where $\eta^{(i)}$ for $i = 1, \ldots, m - \text{rank}(M)$ are arbitrary kernel basis vectors of M. Because no binary variables are introduced in this case, weakly reversible deficiency zero realizations can be computed via pure LP in polynomial time.

The objective function in both of the previous cases can be any linear function of the decision variables. A practical choice can be the minimization of the sum of the off-diagonal elements of A_k. Or, with a slight modification of the constraints, dense or sparse deficiency zero realizations can also be computed.

Example 13. This example taken from [54] illustrates the LP approach for computing weakly reversible deficiency zero structures, and clearly shows that the additional transformation parameters introduced by linear conjugacy might be necessary to find such a reaction graph. The starting kinetic system is given by the matrices Y and M as follows:

$$Y = \begin{bmatrix} 0 & 1 & 2 & 0 & 0 & 0 \\ 0 & 0 & 0 & 1 & 2 & 0 \\ 0 & 0 & 0 & 1 & 0 & 2 \end{bmatrix}, \quad M = \begin{bmatrix} 1 & -1 & -1 & 1 & 0 & 0 \\ 0 & 2 & 0 & -2 & -2 & 2 \\ 0 & 1 & 0 & -1 & 1 & -1 \end{bmatrix} \tag{4.77}$$

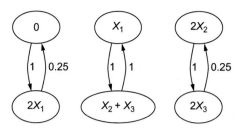

Fig. 4.2 Reaction graph of the obtained linearly conjugate realization of the kinetic system (4.77). This realization is weakly reversible and has zero deficiency.

This representation defines the following ODEs:

$$\dot{x}_1 = 1 - x_1 - x_1^2 + x_2 x_3$$
$$\dot{x}_2 = 2x_1 - 2x_2 x_3 - 2x_2^2 + 2x_3^2 \qquad (4.78)$$
$$\dot{x}_3 = x_1 - x_2 x_3 + x_2^2 - x_3^2$$

It is easy to check that there exists a positive vector in $\ker(M)$ (e.g., $\underline{1}$), so we can apply the LP computation method presented earlier. If we try to find a weakly reversible dynamically equivalent deficiency zero realization, we find that the constraints (4.51)–(4.54) and (4.76) are infeasible with $D = I$.

However, we can find a weakly reversible realization which is linearly conjugate to the original system (4.77) with the transformation $D = \mathrm{diag}\{[2, 1, 2]^T\}$ that has zero deficiency. This CRN is given by (Y, A_k') where the Kirchhoff matrix A_k' has the following nonzero off-diagonal entries: $A'_{k\,3,1} = 1$, $A'_{k\,4,2} = 1$, $A'_{k\,1,3} = 0.25$, $A'_{k\,2,4} = 1$, $A'_{k\,6,5} = 1$, and $A'_{k\,5,6} = 0.25$.

One can see the reversible reaction graph of this network in Fig. 4.2. Therefore, we obtain from the Deficiency Zero Theorem that the kinetic system (4.78) has precisely one at least locally asymptotically stable, strictly positive equilibrium point in each stoichiometric compatibility class.

4.5.3.4 Further Structures Computable in an Optimization Framework

Realizations with several other properties can be computed using LP- or MILP-based optimization such as reversible and detailed balanced networks, or networks with minimal number of complexes [53]. If there is no deficiency zero realization, we are still able to compute weakly reversible realizations with minimal deficiency [57].

4.5.4 Computing All Possible Graph Structures of a Kinetic System

After the special cases treated in this section, the question naturally arises, whether it is possible to enumerate all distinct reaction graph structures realizing a given kinetic dynamics. In this section we will present a solution for this problem based on [55].

Since the aim is to compute distinct graph structures, in this section reaction graphs will be assumed to be unweighted directed graphs (i.e., by reaction graph we will mean only its structure). It is clear from the superstructure property described earlier in Section 4.5.2 that if G_D is the reaction graph of the dense realization, and graph G_R describes another realization, then $E(G_R) \subseteq E(G_D)$ holds, where $E(G)$ denotes the set of edges of the reaction graph G. As it has also been described in Section 4.5.2, the dense realization can be determined by a polynomial-time algorithm, which is the first step of the method presented here.

There might be reactions that take place in each realization. These are called *core reactions*, and the edges representing them are the *core edges*. The set of core edges will be denoted by E_c, which can also be determined by a polynomial algorithm [52]. It is worth doing this step, since it might save significant computational time, however, it is not necessary for the running of the algorithm.

Based on the earlier, each reaction graph can be uniquely determined if it is known which noncore reactions of the dense realization take place in the reaction network. Thus we represent the reaction graphs by binary sequences of length $N = |E(G_D) \setminus E_c|$.

In order to define the binary sequences we fix an ordering of the noncore edges. Let e_i denote the ith edge. If R is a binary sequence, then let $R[i]$ denote the ith coordinate of the sequence and G_R be the reaction graph described by R. If a realization can be encoded by the binary sequence R, then

$$e \in E(G_R) \Longleftrightarrow \begin{cases} e \in E_c \\ \quad \text{or} \\ \exists i \in \{1, \ldots, N\} \quad e = e_i, \; R[i] = 1 \end{cases} \tag{4.79}$$

From now on the term "sequence" will refer to such a binary sequence of length N. The sequence representing the dense realization (with all coordinates equal to 1) will be denoted by D.

For efficient operation of the algorithm, we also need appropriate data structures. The computed graph structures are stored in a binary array of size 2^N called *Exist*, where the indices of the fields are the sequences as binary numbers. At the beginning the values in every field are zero, and after the computation the value of field *Exist*[R] is 1 if and only if there is a linearly conjugate realization described by the sequence R.

We also need $N + 1$ stacks, indexed from 0 to N. The kth stack is referred to as $S(k)$. During the computation, sequences are temporarily stored in these stacks, following the rule: the sequence R describing a linearly conjugate realization might be in stack $S(k)$ if and only if there are exactly k coordinates of R which are equal to 1 (i.e., exactly k reactions take place in the realization).

At the beginning all stacks are empty, but during the running of the algorithm we push in and pop out sequences from them. The command "push R into $S(k)$" pushes the sequence R into the stack $S(k)$, and the command "pop from $S(k)$" pops a sequence out from $S(k)$ and returns it. (It makes no difference, in what order are the elements of the stacks popped out, but by the definition of the data structure the sequence pushed in last will be popped out first.) The number of sequences in stack $S(k)$ are denoted by size.$S(k)$, and the number of coordinates equal to 1 in the sequence R are referred to as $e(R)$.

We will use the following subroutine in the algorithm:

FindLinConjWithoutEdge(M, Y, R, i) computes a constrained dense linearly conjugate realization of the kinetic system with coefficient matrix M and complex composition matrix Y. The additional inputs R and i are a sequence encoding the input reaction graph structure, and an integer index, respectively. The procedure returns a sequence U encoding the graph structure of the computed realization such that G_U is a subgraph of G_R and $U[i] = 0$. (These requirements can easily be expressed by linear constraints.) If there is no such realization, then -1 is returned. This computation can be carried out in polynomial time as it is described in Section 4.5.2.

Now, the pseudocode of the algorithm computing all distinct graph structures of a kinetic system can be given as follows.

Algorithm 2 Determines all reaction graphs describing linearly conjugate realizations of a kinetic system

---

Inputs: M, Y, D

Output: *Exist*

1: push D into $S(N)$
2: *Exist*$[D] := 1$
3: **for** $k = N$ to 1 **do**
4:      **while** size.$S(k) > 0$ **do**
5:          $R := $ pop $S(k)$
6:          **for** $i = 1$ to N **do**
7:              **if** $R[i] = 1$ **then**
8:                  $U := $ FindLinConjWithoutEdge(M, Y, R, i)
9:                  **if then**$U \geq 0$ **and** *Exist*$[U] = 0$
10:                     *Exist*$[U]:= 1$
11:                     push U into $S(e(U))$
12:                 **end if**
13:             **end if**
14:         **end for**
15:         Print R
16:     **end while**
17: **end for**

It is proved in [55] that for any kinetic system and any suitable fixed set of complexes, all the possible reaction graphs describing linearly conjugate realizations can be computed after finitely many steps by Algorithm 2. The whole computation might last until exponential time depending on the number of different reaction graphs, but the time elapsed between the displaying of two linearly conjugate realizations is always polynomial. The algorithm can also be parallelized efficiently.

Example 14. This example was published in [55]. The kinetic system to be studied is given by the following ODEs:

$$\dot{x}_1 = 3k_1 \cdot x_2^3 - k_2 \cdot x_1^3$$
$$\dot{x}_2 = -3k_1 \cdot x_2^3 + k_2 \cdot x_1^3$$

According to the monomials there are (at least) two species, $S = \{X_1, X_2\}$, and we fix the set of complexes to be $C = \{C_1, C_2, C_3\}$, where $C_1 = 3X_2$,

$C_2 = 3X_1$, and $C_3 = 2X_1 + X_2$. Based on the previous description, the Y and M matrices are the following:

$$Y = \begin{bmatrix} 0 & 3 & 2 \\ 3 & 0 & 1 \end{bmatrix} \quad M = \begin{bmatrix} 3k_1 & -k_2 & 0 \\ -3k_1 & k_2 & 0 \end{bmatrix} \tag{4.80}$$

Moreover, the structure of the dense realization computed by Algorithm 1 is a fully connected directed graph (depicted as G_1 in Fig. 4.3).

For the numerical computations, the parameter values $k_1 = 1$ and $k_2 = 2$ were used. As the result of the algorithm we obtained 18 different sequences/reaction graphs. This small example is special in the sense that the sets of different reaction graphs corresponding to dynamically equivalent and linearly conjugate realizations are the same, since the computed transformation matrix D was the unit matrix in each case. Using the numerical results, it was easy to symbolically solve the equations for dynamical equivalence; therefore, we can give the computed reaction rate coefficients as functions of k_1 and k_2.

The reaction graphs are denoted by $G_1, \ldots, G_{18}$ and are presented with suitable reaction rate coefficients in Fig. 4.3. From the computation it follows that there are two reaction rate coefficients k_{31} and k_{32} which do not depend on the input parameters, just on each other, and the reactions determined by these might together be present or nonpresent in the reaction network. Therefore, a nonnegative parameter p is applied to determine the values of these coefficients. We get the reaction graphs $G_1, \ldots, G_9$ if the parameter p is positive, and if it is 0, then we get the reaction graphs $G_{10}, \ldots, G_{18}$. The reaction graph G_1 (the complete directed graph) describes the dense realization, and consequently, all other reaction graphs are subgraphs of it (not considering the edge weights).

Example 15. Let us revisit again the model of the irreversible Michaelis-Menten kinetics studied in Examples 5 and 6. If we fix the complex set to be $C_1 = X_1 + X_2$, $C_2 = X_3$, $C_3 = X_2 + X_4$, then we find that the reaction graph shown in Fig. 2.1 is unique, since the dense and sparse structures coincide. This means that all three reactions of the system are core reactions.

Now, let us define the complex set as the union of the complexes appearing in Examples 5 and 6, that is,

$$C_1 = X_1 + X_2, \ C_2 = X_3, \ C_3 = X_2, \ C_4 = X_1, \ C_5 = X_1 + X_2 + X_3$$
$$C_6 = X_1 + X_3, \ C_7 = X_2 + X_3, \ C_8 = 0, \ C_9 = X_3 + X_4, \ C_{10} = X_2 + X_4 \tag{4.81}$$

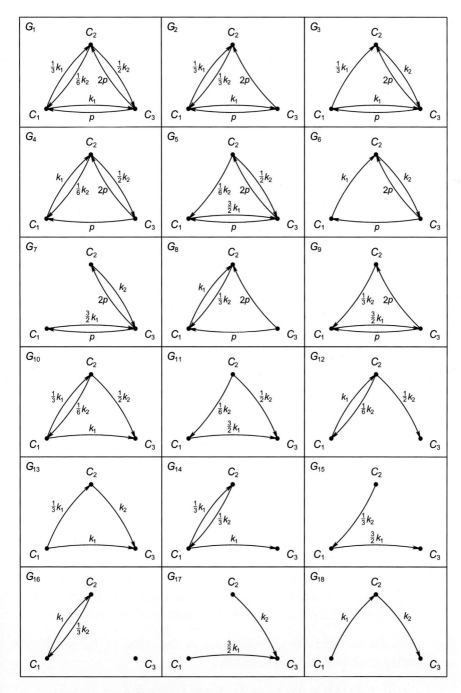

Fig. 4.3 All reaction graphs of the kinetic system (4.80) with possible reaction rate coefficients.

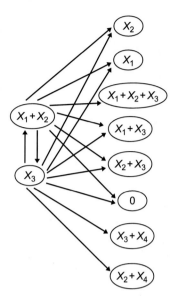

Fig. 4.4 *Structure of the dense realization of the kinetic system defined by Eq. (2.51) using the complex set listed in Eq. (4.81).*

The parameter values for the computations were $k_1^+ = 2$, $k_1^- = 1$, and $k_2^+ = 0.1$.

The structure of the dense realization is shown in Fig. 4.4. It is visible that the reaction graphs shown in Figs. 2.1 and 2.2 are indeed the subgraphs of the dense realization. It is also interesting that we find no core reactions in this case.

4.5.5 Computation of Linearly Conjugate Bio-CRNs

This section is based on [26]. It is easy to see that Eq. (3.16) is again a standard linear constraint for the elements of A_b and D, because it can be rewritten as

$$DM - YA_b = 0 \qquad (4.82)$$

Let us use the following notation

$$D = \text{diag}(\tilde{c}) \qquad (4.83)$$

In additional, let us denote the elements of the Kirchhoff matrix $A_{b\,ij}$ by a_{ij} for $i = 1, \ldots, m, j = 1, \ldots, \kappa$. Then Eq. (4.82) with the positivity constraint on D can be written as

$$\tilde{c}_i > 0 \quad \text{for} \quad i = 1, \ldots, n$$

$$\tilde{c}_i M_{ij} - \sum_{k=1}^{m} Y_{ik} a_{kj} = 0 \quad \text{for} \quad i = 1, \ldots, n, \ j = 1, \ldots, \kappa \qquad (4.84)$$

We know that A_b is a column conservation matrix. This requirement can be formulated as additional linear equality constraints. For convenience, these constraints are given for each block ($p = 1, \ldots, m$) of the Kirchhoff matrix (2.57) as follows:

$$a_{d_p j} = - \sum_{i=1, \ i \neq d_p}^{m} a_{ij} \quad \text{for} \quad j = 1, \ldots, d_p, \ p = 1, \ldots, m \qquad (4.85)$$

Eqs. (4.84), (4.85) represent $n \times \kappa + \kappa$ equality constraints to find $n + m \times \kappa$ variables.

For practical reasons, it might be required to consider bound constraints on the continuous variables as

$$a_{ij}^{lb} \leq a_{ij} \leq a_{ij}^{ub} \quad \text{for} \ i = 1, \ldots, m, \ j = 1, \ldots, \kappa$$
$$\tilde{c}_i^{lb} \leq \tilde{c}_i \leq \tilde{c}_i^{ub} \quad \text{for} \ i = 1, \ldots, n \qquad (4.86)$$

where $a_{ij}^{lb}, a_{ij}^{ub}, \tilde{a}_{ij}^{lb}, \tilde{c}_{ij}^{ub}$ represent lower and upper bounds for the entries of the Kirchhoff matrix and for the scaling parameters, respectively.

The prescription of additional properties such as density/sparsity, or the enumeration of all possible graph structures is possible in the case of bio-CRNs as well with the appropriate adaptation of the methods described in Sections 4.5.2–4.5.4.

4.5.6 Computation-Oriented Representation of Uncertain Kinetic Models and Their Realizations

Even if the monomials of a kinetic system are known, the parameters (i.e., the monomial coefficients) are often uncertain in practice. For example, one may consider the situation when the rate coefficients of a kinetic polynomial ODE model with fixed structure are identified from noisy measurement data. In such a case, an interval-based or more generally, a polytopic uncertain model can be constructed based on the uncertainty of the estimate (see Section B.4.5 for more details). The results summarized in this section are mainly taken from [58], where the detailed proofs can be found.

4.5.6.1 Representation of the Uncertainty in Kinetic Models

Let us consider again the polynomial ODE (4.50). For the uncertainty modeling, we assume that the monomial coefficients in matrix M are constant but uncertain, and they belong to an $n \cdot m$ dimensional polyhedron $\mathcal{P}$ (see Section B.4.5 for explanation).

We represent the matrix M as a point denoted by $\tilde{M}$, in the Euclidean space $\mathbb{R}^{nm}$. In the uncertain model it is assumed that the possible points $\tilde{M}$ are all the points of a closed convex polyhedron $\mathcal{P}$, which is defined as the intersection of q halfspaces. The boundaries of the halfspaces are hyperplanes with normal vectors $n_1, \ldots, n_q \in \mathbb{R}^{nm}$ and constants $b_1, \ldots, b_q \in \mathbb{R}$. Applying these notations, the polyhedron $\mathcal{P}$ can be described by a set of linear inequalities (see also Eq. B.31) as

$$\mathcal{P} = \{\tilde{M} \in \mathbb{R}^{nm} \mid \tilde{M}^\top \cdot n_i \leq b_i, \ 1 \leq i \leq q\} \tag{4.87}$$

For the characterization of the polyhedron $\mathcal{P}$, one should consider not only the possible values of the parameters, but also the kinetic property of the polynomial system. This can be ensured (see [21]) by prescribing the sign pattern of the matrix M as follows:

$$Y_{ij} = 0 \implies M_{ij} \geq 0, \quad i \in \{1, \ldots, n\}, j \in \{1, \ldots, m\} \tag{4.88}$$

These constraints are of the same form as the inequalities in Eq. (4.87). For example, the constraint $\tilde{M}_j \geq 0$ can be written by choosing the normal vector n_i to be the unit vector $-e_j^{nm}$ and b_i to be the null vector $\underline{0}^{nm}$.

It is possible to define a set L of finitely many additional linear constraints on the variables to characterize a special property of the realizations, for example, a set of reactions to be excluded, or mass conservation on a given level (e.g., see [55]). These constraints can affect not only the entries of the coefficient matrix M but also the Kirchhoff matrix of the realizations. If the Kirchhoff matrix A_k of the realization is represented by the point $\tilde{A}_k \in \mathbb{R}^{m^2 - m}$ storing the off-diagonal elements, and r is the number of constraints in the set L, then the equations can be written in the form

$$\tilde{M}^\top \cdot \alpha_i + \tilde{A}_k^\top \cdot \beta_i \leq d_i \tag{4.89}$$

where $\alpha_i \in \mathbb{R}^{nm}$, $\beta_i \in \mathbb{R}^{m^2 - m}$, and $d_i \in \mathbb{R}$ hold for all $i \in \{1, \ldots, r\}$. These constraints can be modeled in the framework of linear programming.

4.5.6.2 Realizations of an Uncertain Kinetic Model

In the case of the uncertain model, we will examine realizations assuming a fixed set of complexes. Therefore, the known parameters are the polyhedron

$\mathcal{P}$, the set L of constraints, and the matrix Y. Hence, an *uncertain kinetic system* is referred to as the triple $[\mathcal{P}, L, Y]$.

Definition 1. A reaction network (Y, A_k) is called a *realization of the uncertain kinetic system* $[\mathcal{P}, L, Y]$ if there exists a coefficient matrix $M \in \mathbb{R}^{n \times m}$ so that the equation $M = Y \cdot A_k$ holds, the point $\tilde{M}$ is in the polyhedron $\mathcal{P}$, and the entries of the matrices M and A_k fulfill the set L of constraints. Since the matrix Y is fixed but the coefficients of the polynomial system can vary, this realization is referred to as the matrix pair (M, A_k).

Assuming a fixed set of complexes, a realization (M, A_k) of an uncertain kinetic system $[\mathcal{P}, L, Y]$ can be computed using a linear optimization framework.

In the constraint satisfaction or optimization model, the decision variables are the entries of the matrix M and the off-diagonal entries of the matrix A_k. The constraints regarding the realizations of the uncertain model can be written as follows:

$$\tilde{M}^\top \cdot n_i \leq b_i, \quad i \in \{1, \ldots, p\} \tag{4.90}$$

$$M = Y \cdot A_k \tag{4.91}$$

$$A_{k\,ij} \geq 0, \quad i \neq j, \; i,j \in \{1, \ldots, m\} \tag{4.92}$$

$$\sum_{j=1}^{m} A_{k\,ij} = 0, \quad j \in \{1, \ldots, m\} \tag{4.93}$$

Eq. (4.90) ensures that the parameters of the dynamics correspond to a point of the polyhedron $\mathcal{P}$. Dynamical equivalence is defined by Eq. (4.91), while Eqs. (4.92), (4.93) are required for the Kirchhoff property of matrix A_k to be fulfilled. Moreover, the constraints in the set L can be written in the form of Eq. (4.89).

The objective function of the optimization model can be defined according to the desired properties of the realization, for example, in order to examine if the reaction $C_i \rightarrow C_j$ can be present in the reaction network or not, the objective can be defined as $\max[A_k]_{ji}$.

We apply the representation of realizations of the uncertain model as points of the Euclidean space $\mathbb{R}^{m^2-m+nm}$. The coordinates with indices $i \in \{1, \ldots, m^2 - m\}$ characterize the Kirchhoff matrix of the realization and the remaining coordinates $j \in \{m^2 - m + 1, \ldots, m^2 - m + nm\}$ define the coefficient matrix M of the polynomial system. Due to the linearity of the constraints in the computational model, the set of possible realizations of an uncertain kinetic system $[\mathcal{P}, L, Y]$ is a convex bounded polyhedron denoted by $\mathcal{Q}$.

4.5.6.3 Dense and Sparse Realizations
Dense and sparse realizations of an uncertain kinetic model, which are useful during the structural analysis, can be introduced in the case of the uncertain model as well.

Definition 2. A realization (M, A_k) of the uncertain kinetic system $[\mathcal{P}, L, Y]$ is called a *dense (sparse)* realization if it has the maximum (minimum) number of reactions.

It can be proved that the superstructure property holds for uncertain kinetic systems as well, and the proof is based on the same idea as in the nonuncertain case, see [51].

Proposition 3. *A dense realization (M, A_k) of an uncertain kinetic system $[\mathcal{P}, L, Y]$ determines a superstructure among all realizations of the model.*

It follows from Proposition 3 that the structure of the dense realization is unique. If there were two different dense realizations, then the reaction graphs representing them would contain each other as subgraphs, which implies that these graphs are structurally identical.

The dense and sparse realizations are useful for checking the structural uniqueness of the uncertain model.

Proposition 4. *The dense and sparse realizations of an uncertain kinetic system $[\mathcal{P}, L, Y]$ have the same number of reactions if and only if all realizations of the model are structurally identical.*

A dense realization of the uncertain kinetic system can be computed by the application of a recursive polynomial-time algorithm. The basic principle of the method is the same as for Algorithm 1.

The computation can be performed in polynomial time, too, since it requires at most $m^2 - m$ steps of LP optimization and some minor computation.

In the algorithm the assigned realizations are represented as points in $\mathbb{R}^{m^2 - m + nm}$ and are determined using the following procedure.

FindPositive$([\mathcal{P}, L, Y], H)$ returns a pair (R, B). The point $R \in \mathcal{Q}$ represents a realization of the uncertain model $[\mathcal{P}, L, Y]$ for which the value of the objective function $\sum_{j \in H} R_j$ considering a set $H \subseteq \{1, \ldots, m^2 - m\}$ of indices is maximal. The other returned object is a set B of indices where $k \in B$ if and only if $Q_k > 0$. If there is no realization fulfilling the constraints then the pair $(\mathbf{0}, \emptyset)$ is returned.

In the algorithm we apply the arithmetic mean as convex combination (i.e., if the number of the assigned realizations is k then all the coefficients of the convex combination are $\frac{1}{k}$).

Algorithm 3   Computes a dense realization for an uncertain kinetic model

Input: $[\mathcal{P}, L, Y]$
Output: *Result*
 1: $H := \{1, \ldots, m^2 - m\}$
 2: $B := H$
 3: *Result* $:= \mathbf{0} \in \mathbb{R}^{m^2 - m + nm}$
 4: *loops* $:= 0$
 5: **while** $B \neq \emptyset$ **do**
 6:      $(R, B) :=$ FindPositive$([\mathcal{P}, L, Y], H)$
 7:      *Result* $:=$ *Result* $+ R$
 8:      $H := H \setminus B$
 9:      *loops* $:=$ *loops* $+ 1$
10: **end while**
11: *Result* $:=$ *Result*$/loops$
12: **if** *Result* $= \mathbf{0}$ **then**
13:      There is no realization with the given properties.
14: **else**
15:      *Result* is a dense realization.
16: **end if**

Proposition 5. *The realization returned by Algorithm 3 is a dense realization of the uncertain kinetic system.*

Stabilizing Feedback Control Design

We consider positive polynomial systems extended with polynomial input-affine input terms in this chapter that can be written in the form of an ODE

$$\frac{dx}{dt} = \mathcal{P}_s(x) + \mathcal{P}_u(x) \cdot u \tag{5.1}$$

where $x \in \overline{\mathbb{R}}_+^n$ is the state variable, $u \in \overline{\mathbb{R}}_+^p$ is the vector of input variables, and $\mathcal{P}_s$, $\mathcal{P}_u$ are polynomial or quasipolynomial (QP) functions.

A special but important control aim, the stabilization of the previous systems is only considered here, that is achieved by applying a static

Analysis and Control of Polynomial Dynamic Models with Biological Applications.
https://doi.org/10.1016/B978-0-12-815495-3.00005-5

polynomial or QP state feedback in the form

$$u = \mathcal{P}_c(x) \tag{5.2}$$

This way the closed-loop system

$$\frac{dx}{dt} = \mathcal{P}_s(x) + \mathcal{P}_u(x) \cdot \mathcal{P}_c(x) \tag{5.3}$$

will also be polynomial or QP.

5.1 STABILIZING CONTROL OF QP SYSTEMS BY USING OPTIMIZATION

In Section 2.2 we have already introduced the notion of QP systems, where the state and input mappings of Eq. (5.1) $\mathcal{P}_s(x)$ and $\mathcal{P}_u(x)$ are both QP mappings. In order to have the closed-loop system (5.3) also in QP form, the nonlinear mapping in the feedback $\mathcal{P}_c(x)$ is also chosen to be QP.

5.1.1 LQ Control of QP Systems Based on Their Locally Linearized Dynamics

The aim of this section is to apply an *LQ-based state feedback controller* for QP and Lotka-Volterra (LV) systems through a locally linearized model corresponding to a (unique) positive equilibrium point of the closed-loop system. The primary aim is the formulation of an LQ problem that yields a diagonally stable LTI system. This implies that the corresponding LV or QP system will be globally asymptotically stable.

Let us consider a *linear input structure* for the original QP model in Section 2.2, that can be formally derived by regarding λ as a function of the input vector u

$$\lambda = \phi u, \quad u \in \mathbb{R}^r, \quad \phi \in \mathbb{R}^{n \times r}, \quad r \le n$$

such that the state equation extended with an input term is in the form

$$\frac{dx}{dt} = \operatorname{diag}\{x\} \left(\mathcal{A} p + \phi u\right) \tag{5.4}$$

where $p_j = \prod_{k=1}^n x_k^{B_{j,k}}$ for $j = 1, \dots, m$ are the quasimonomials.

5.1.1.1 The Equivalence of the Diagonal Stability of the QP and Linearized-QP Models

The linearized version of the QP model (5.4) assuming $u \equiv 0$ around its positive equilibrium point x^* is in the form

$$\frac{d\Delta x}{dt} = \left[\operatorname{diag}\{x^*\} \, \mathcal{A} \operatorname{diag}\{p^*\} \, \mathcal{B} \operatorname{diag}\{x^*\}^{-1} \right] \Delta x \qquad (5.5)$$

where $\Delta x = x - x^*$. When the matrix $\mathcal{B}$ is invertible (i.e., we have the same number m of quasimonomials as the number of state variables n, $m = n$), then we can transform (5.5) with the linear transformation

$$x' = \left[\operatorname{diag}\{p^*\} \, \mathcal{B} \operatorname{diag}\{x^*\}^{-1} \right] \Delta x = T \Delta x$$

with the transformation matrix

$$T = \operatorname{diag}\{p^*\} \, \mathcal{B} \operatorname{diag}\{x^*\}^{-1} \qquad (5.6)$$

and the transformed system is in the form

$$\frac{dx'}{dt} = \operatorname{diag}\{p^*\} \, \mathcal{B} \, \mathcal{A} x' = \mathcal{M}^* x' \qquad (5.7)$$

The *equivalence of the diagonal stability of the original QP and the transformed linearized system (5.7) can be easily shown*, see [59].

5.1.1.2 LQ Feedback Structure

The proposed feedback structure may achieve diagonal stability and improves the local LQ performance, but one *must assume an invertible (i.e., square and full rank) quasimonomial composition matrix $\mathcal{B}$.*

Let us consider the nonlinear state feedback in the form

$$u = Kp + u^* \qquad (5.8)$$

where p is a vector of quasimonomials, and u^* is a constant that moves the closed-loop equilibrium to the desired steady-state point x^*.

Let us consider the steady-state equations in the closed-loop case

$$\phi \, (Kp^* + u^*) + \mathcal{A} q^* = 0 \qquad (5.9)$$

Let

$$u^* = -Kp^* + u^{**} \qquad (5.10)$$

then we can set the positive vector p^* to a steady state of the closed-loop system if there exists u^{**} such that

$$\phi \, u^{**} + \mathcal{A} p^* = 0 \qquad (5.11)$$

When the positive vector x^* is a suitable steady state of the closed-loop system, then the linearized closed-loop QP system is

$$\frac{d\Delta x}{dt} = \left(\text{diag}\{x^*\} \left[\mathcal{A} + \phi K \right] \text{diag}\{p^*\} \mathcal{B} \, \text{diag}\{x^*\}^{-1} \right) \Delta x \qquad (5.12)$$

In that case the linearized transformed input is

$$u' = T \Delta u = K \, \text{diag}\{p^*\} \mathcal{B} \, \text{diag}\{x^*\}^{-1} \Delta x = K x' \qquad (5.13)$$

5.1.1.3 Suboptimal LQ With Diagonal Stability

With the previous feedback structure, we formulate and solve an LQ control problem with diagonal stability for QP systems. For this, *we assume that* $\mathcal{B}$ *is invertible, that is, it is a full rank square matrix with* $m = n$.

When the matrix $\mathcal{B}$ is invertible, then the linearized closed-loop system (5.12) can be transformed into the following form

$$\dot{x} = (\mathcal{M}^* + \mathcal{N}^* K) x \qquad (5.14)$$

using the transformation matrix T in Eq. (5.6), where $\mathcal{M}^* = \text{diag}(p^*) \mathcal{B} \mathcal{A}$ and $\mathcal{N}^* = \text{diag}(p^*) \mathcal{B} \phi$.

As stated before, when the state matrix of the linearized and transformed closed-loop system

$$\mathcal{M}^* + \mathcal{N}^* K \qquad (5.15)$$

is diagonally stable, then the corresponding QP system is diagonally stable, too.

If the pair $(\mathcal{M}^*, \mathcal{N}^*)$ is diagonally stabilizable, then a suboptimal LQ of the linearized closed-loop system with diagonal stability can be computed by following the method proposed in [11] as follows.

Let the LQ objective function of the closed-loop linearized QP system be

$$J(K) = \int_0^\infty \|Q^{1/2} \Delta x\|_2^2 + \|R^{1/2} K \, \text{diag}(q^*) \mathcal{B} \, \text{diag}(x^*)^{-1} \Delta x\|_2^2 \, dt \quad (5.16)$$

where Q and R are positive definite matrices with the appropriate dimensions, that can be chosen arbitrarily to satisfy our control performance aims. To guarantee diagonal stability, we have to design the feedback to the transformed and linearized system. Then, the transformed LQ objective is in the form

$$J(K) = \int_0^\infty \|Q^{1/2}T^{-1}x'\|_2^2 + \|R^{1/2}Kx'\|_2^2 \, dt \qquad (5.17)$$

using the transformation matrix T in Eq. (5.6). The suboptimal feedback gain K can be computed by solving the following LMI problem

$$\min_{P,X,Y} \mathrm{Tr}(T^{-T}QT^{-1}P) + \mathrm{Tr}(X) \qquad (5.18)$$

subject to

$$\mathcal{M}^* P + P(\mathcal{M}^*)^T + N^* Y + Y^T (N^*)^T + I < 0 \qquad (5.19)$$

$$\begin{bmatrix} X & R^{1/2}Y \\ Y^T R^{1/2} & P \end{bmatrix} > 0 \qquad (5.20)$$

where P is a *positive diagonal matrix*, X is a positive definite matrix, and $Y = KP$.

The resulting closed-loop system (5.14) is diagonally stable and has suboptimal LQ performance such that

$$J(K_{\mathrm{opt}}) \leq J(K) \leq \mathrm{Tr}(T^{-T}QT^{-1}P) + \mathrm{Tr}(X) \qquad (5.21)$$

where K and K_{opt} are the solution of the problems (5.18)–(5.20) with and without the diagonal restriction, respectively.

5.1.2 Stabilizing Control of QP Systems by Solving Bilinear Matrix Inequalities

The input-affine extended form of the QP model in Section 2.2 is considered here in the form of the following ODE

$$\frac{dx_i}{dt} = x_i \left(\lambda_{0i} + \sum_{j=1}^m A_{0i,j} \prod_{k=1}^n x_k^{B_{j,k}} \right) + \sum_{\ell=1}^r x_i \left(\lambda_{\ell i} + \sum_{j=1}^m A_{\ell i,j} \prod_{k=1}^n x_k^{B_{j,k}} \right) \cdot u_\ell,$$

$$i = 1, \ldots, n \qquad (5.22)$$

where the vector of state variables x and that of the input variables u are defined on the nonnegative orthant, and $p_j = \prod_{k=1}^n x_k^{B_{j,k}}$ for $j = 1, \ldots, m$ are the quasimonomials.

5.1.2.1 Controller Design Problem

The globally stabilizing QP state feedback design problem for QP systems can be formulated as follows. Consider arbitrary QP inputs in the form

$$u_\ell = \sum_{i=1}^{\rho} k_{i\ell} \hat{p}_i, \quad \ell = 1 \ldots, r \tag{5.23}$$

where $\hat{p}_i = \hat{p}_i(x_1, \ldots, x_n)$, $i = 1, \ldots, r$ are arbitrary quasimonomial functions of the state variables of Eq. (5.22) and $k_{i\ell}$ is the constant gain of the quasimonomial function $\hat{p}_i$ in the ℓth input u_ℓ. The closed-loop (autonomous) system will also be a QP system with matrices

$$\hat{A} = A_0 + \sum_{\ell=1}^{r} \sum_{i=1}^{\rho} k_{i\ell} A_\ell, \quad \hat{B} \tag{5.24}$$

$$\hat{\lambda} = \lambda_0 + \sum_{\ell=1}^{r} \sum_{i=1}^{\rho} k_{i\ell} \lambda_\ell \tag{5.25}$$

Note that the number of quasimonomials in the closed-loop system (i.e., the dimension of the matrices) together with the matrix $\hat{B}$ may significantly change depending on the choice of the feedback structure, that is, on the quasimonomial functions $\hat{p}_i$.

Furthermore, the closed-loop LV coefficient matrix $\hat{\mathcal{M}}$ can also be expressed in the form

$$\hat{\mathcal{M}} = \hat{B} \cdot \hat{A} = \mathcal{M}_0 + \sum_{\ell=1}^{r} \sum_{i=1}^{\rho} k_{i\ell} \mathcal{M}_\ell$$

Then the global stability analysis of the closed-loop system with unknown feedback gains $k_{i\ell}$ leads to the following bilinear matrix inequality (BMI)

$$\hat{\mathcal{M}}^T C + C\hat{\mathcal{M}} = \mathcal{M}_0^T C + C\mathcal{M}_0 + \sum_{\ell=1}^{r} \sum_{i=1}^{\rho} k_{i\ell} \left(\mathcal{M}_\ell^T C + C\mathcal{M}_\ell \right) \leq 0 \tag{5.26}$$

The variables of the BMI are the $r \times \rho$ $k_{i\ell}$ feedback gain parameters and the $c_j, j = 1, \ldots, m$ parameters of the Lyapunov function. If the BMI above is feasible, then there exists a globally stabilizing feedback with the selected structure.

5.1.2.2 Numerical Solution of the Controller Design Problem

There are just a few software tools available for solving general bilinear matrix inequalities that is a computationally hard problem. In some rare fortunate cases, with a suitable change of variables, quadratic matrix inequalities can be rewritten as linear matrix inequalities (see, e.g., [38]). Unfortunately, the structure of the matrix variable of Eq. (5.26) does not fall into this fortunate problem class, so the previously mentioned idea cannot be used.

Rewriting the previous matrix inequality (5.26) in the form of a basic BMI problem, one gets the following expression which can be directly solved by Kocvara and Stingl [60] as a BMI feasibility problem:

$$\sum_{j=1}^{m} c_j \bar{M}_{0,j} + \sum_{j=1}^{m} \sum_{l=1}^{p} \sum_{i=1}^{r} c_j k_{il} \bar{M}_{il,j} \leq 0$$

$$-c_1 < 0 \qquad\qquad (5.27)$$

$$\vdots$$

$$-c_m < 0$$

The two disjoint sets of BMI variables are the c_j parameters of the Lyapunov function and the $k_{i\ell}$ feedback parameters. The parameters of the problem $\bar{M}_{0,j}$ ($\bar{M}_{il,j}$, respectively) are the symmetric matrices obtained from $\mathcal{M}_0$ ($\mathcal{M}_\ell$, respectively) by adding the $m \times m$ matrix that contains only the jth column of $\mathcal{M}_0$ ($\mathcal{M}_\ell$, respectively) to its transpose:

$$\mathcal{M} = \begin{bmatrix} M_{11} & \cdots & M_{1j} & \cdots & M_{1m} \\ \vdots & \ddots & \vdots & \ddots & \vdots \\ M_{m1} & \cdots & M_{mj} & \cdots & M_{mm} \end{bmatrix}$$

$$\downarrow$$

$$\bar{\mathcal{M}}_j = \begin{bmatrix} 0 & \cdots & M_{1j} & \cdots & 0 \\ \vdots & & \vdots & & \vdots \\ M_{1j} & \cdots & 2M_{jj} & \cdots & M_{mj} \\ \vdots & & \vdots & & \vdots \\ 0 & \cdots & M_{mj} & \cdots & 0 \end{bmatrix}$$

Because of the NP-hard nature of the general BMI solution problem, it is worthwhile to search for an approximate, but numerically efficient, alternative way of solution. It was shown in [34] that the special structure of the QP stabilizing feedback design BMI feasibility problem allows us to apply a computationally feasible method for its solution that solves an LMI in each of its iterative approximation step. The already existing iterative

LMI (ILMI) algorithm of [61] used for static output feedback stabilization (see, e.g., in [61]) was used for this purpose.

Example 16 (A simple QP model). Consider the following open generalized mass-action law system

$$\begin{aligned}
\dot{x}_1 &= 0.5x_1 - x_1^{2.25} - 0.5x_1^{1.5}x_2^{0.25} + u_1 \\
\dot{x}_2 &= x_2 - 0.5x_2^{1.75} + u_2
\end{aligned} \tag{5.28}$$

where x_1 and x_2 are the concentrations of chemical species X_1 and X_2 (moles/m^3), while u_1 and u_2 (the manipulable inputs) are their volume-specific component mass inflow rates (moles/m^3 s). The previous two differential equations originate from the component mass conservation equations constructed for a perfectly stirred balance volume [62] under the following modeling assumptions:

1. constant temperature and overall mass,
2. constant physicochemical properties (e.g., density),
3. presence of an inert solvent in a great excess,
4. presence of the following reaction network:
   - autocatalytic generation of the species X_1 and X_2 (e.g., by polymer degradation when they are the monomers and the polymers are present in a great excess) giving rise to the reaction rates $0.5x_1$ and x_2 (the first terms in the right-hand sides), respectively,
   - a self-degradation of these species described by the reaction rates $-x_1^{2.25}$ and $-0.5x_2^{1.75}$ (the second terms on the right-hand sides), respectively,
   - a catalytic degradation of the specie X_1 catalyzed by specie X_2 that corresponds to $-0.5x_1^{1.5}x_2^{0.25}$ in the first equation only (the third term).

The control aim is to drive the system to a positive equilibrium

$$x_1^* = 2.4082 \text{ moles/m}^3, \quad x_2^* = 16.3181 \text{ moles/m}^3$$

This goal can be achieved, for example, by the following nonlinear feedback:

$$\begin{aligned}
u_1 &= 0.5x_1x_2^{0.75} \\
u_2 &= 0.5x_1^{1.25}x_2 + 0.5x_1^{0.5}x_2^{1.25}
\end{aligned} \tag{5.29}$$

The previous inputs, being the component mass flow rates fed to the system (they are both positive), are needed for compensating for the degradation of the specie X_1 and X_2.

By substituting Eq. (5.29) into Eq. (5.28), we obtain the controlled closed-loop system that is a QP system with the following matrices

$$\mathcal{A} = \begin{bmatrix} -1 & 0.5 & -0.5 \\ 0.5 & -0.5 & 0.5 \end{bmatrix} \tag{5.30}$$

$$\mathcal{B} = \begin{bmatrix} 1.25 & 0 \\ 0 & 0.75 \\ 0.5 & 0.25 \end{bmatrix}, \quad \lambda = \begin{bmatrix} 0.5 \\ 1 \end{bmatrix} \tag{5.31}$$

The eigenvalues of the Jacobian matrix of the system at the equilibrium point are -6.4076 and -0.7768.

Since the rank of $\mathcal{M} = \mathcal{B} \cdot \mathcal{A}$ in this case is only 2, we can only use the algorithm described in [63] to prove that the LMI (4.6) is not feasible in this case.

However, by solving Eq. (4.17) using again the algorithm described in [40] we find that we can use the following time reparametrization:

$$\omega = \begin{bmatrix} -0.25 & -0.5 \end{bmatrix}^T \tag{5.32}$$

and the diagonal matrix containing the coefficients of the Lyapunov function is

$$C = \text{diag}([1 \; 2 \; 2 \; 2]) \tag{5.33}$$

The eigenvalues of $\tilde{\mathcal{M}}^T \cdot C + C \cdot \tilde{\mathcal{M}}$ in this case are

$$\lambda_1 = 0, \quad \lambda_2 = 0, \quad \lambda_3 = -4.5, \quad \lambda_4 = -2.5 \tag{5.34}$$

which again proves the global stability of the system.

The previous example demonstrates how time reparametrization can be used in the design of suitable globally stabilizing static feedbacks for nonlinear process systems.

5.2 STABILIZING STATE FEEDBACK CONTROL OF NONNEGATIVE POLYNOMIAL SYSTEMS USING SPECIAL CRN REALIZATIONS OF THE CLOSED-LOOP SYSTEM

We assume for the feedback design that the equations of the *open-loop polynomial system with linear input structure* are given as

$$\dot{x} = M_p \cdot \psi(x) + B \cdot u \tag{5.35}$$

where $x \in \mathbb{R}^n$ is the state vector, $u \in \mathbb{R}^r$ is the input, $\psi \in \mathbb{R}^n \to \mathbb{R}^m$ contains the monomials of the open-loop system, $M_p \in \mathbb{R}^{n \times m}$ and $B \in \mathbb{R}^{n \times r}$.

5.2.1 Controller Design Problem

A polynomial dynamic feedback of the form below is considered for the stabilizing feedback design

$$u = K \cdot \psi(x) \tag{5.36}$$

With this state feedback, the equations of the closed-loop system are given by

$$\dot{x} = \left(M_p + B \cdot K\right) \cdot \psi(x) = M \cdot \psi(x) \tag{5.37}$$

The aim is to *set the closed-loop coefficient matrix $M = (M_p + BK)$ by choosing a suitable feedback gain K such that it defines a kinetic system with prescribed properties (e.g., complex balance) such that the overall complex composition matrix Y is compatible with ψ.*

More precisely, the aim of the feedback is to set a region in the state space, where $x^* \in \overline{\mathbb{R}}_+^n$ is an (at least) locally asymptotically stable equilibrium point of the closed-loop system. This will be achieved in two different ways:

(i)   to find a feedback gain K such that the closed-loop system has a MAL-CRN complex balanced realization with a given equilibrium point x^* and

(ii) to compute the feedback gain K such that the closed-loop system is weakly reversible with deficiency zero.

It is important to note that, as shown in [27],

- it is not necessary to apply dynamic feedback to achieve complex balance or weak reversibility with zero deficiency for the closed-loop system and
- it is sufficient to use only the monomials of the open-loop system (5.35) in the monomial function $\psi(x)$ of the static feedback (5.36).

5.2.1.1 Underlying Realization Computation Problem

It is clear from Section 2.3 that this is possible if and only if the closed-loop coefficient matrix M can be factorized as $M = Y \cdot A_k$ where $Y \in \overline{\mathbb{N}}_+^{n \times m}$ is the given complex composition matrix (which generates $\psi(x)$ in Eq. 5.37), and $A_k \in \mathbb{R}^{m \times m}$ is a valid Kirchhoff matrix. This way one can formulate and solve the feedback design problem as an optimization problem for computing CRN realizations of the closed-loop system.

In the following, two realization computation problems are proposed for dynamically equivalent network structures that solve the feedback gain computation problem for the previous two cases. It will be shown in the following that one can apply linear programming (LP or MILP) optimization approaches to compute the feedback gain K.

5.2.1.2 Constraints on the Feedback Gain

Besides the previous general feedback design problem statement, the proposed linear programming-based approach is capable of handling additional *constraints on the feedback gain*. These can serve to achieve a physically realizable feedback (in case of a nonnegativity requirement on the inputs) or to find a feedback with low gain (when minimizing the l_1 norm).

- *Nonnegativity of the feedback*
  In many cases nonnegativity of the input is a necessary condition of physical realizability. When the closed-loop system is kinetic then its states are nonnegative. Therefore, the feedback $u = K\psi(x)$ is nonnegative if the elements of the matrix K are nonnegative

$$K_{ij} \geq 0, \quad \forall i = 1, \ldots, n, \ j = 1, \ldots, m \qquad (5.38)$$

- *Minimization of the l_1-norm of the feedback gain*
  A suitable objective function of the optimization is the l_1-norm of the matrix K

$$f_{\text{obj}} = \sum_{i=1}^{n} \sum_{j=1}^{m} |K_{ij}| \qquad (5.39)$$

The minimization of this objective function can be easily implemented in the LP framework.

5.2.2 Feedback Computation in the Complex Balanced Closed-Loop Case

In this case, the equilibrium point x^* of the closed-loop system is a design parameter, so it is assumed to be known before the optimization. A complex balanced equilibrium point is (at least) locally asymptotically stable in its stoichiometric compatibility class. Therefore, the aim of the feedback is to set the vector x^* to be a complex balanced equilibrium point of the controlled system.

The linear optimization problem to be solved is described in terms of the linear constraints that express the requirements in the feedback design.

The *first set of constraints is responsible for the dynamical equivalence*

$$\begin{cases} M_p + BK = Y \cdot A_k \\ \underline{1}^T \cdot A_k = \underline{0}^T \\ A_{k\,ij} \geq 0 \quad i,j = 1,\ldots,m, \; i \neq j \end{cases} \tag{5.40}$$

where the elements of K and A_k are the continuous decision variables of the optimization problem, and $\underline{0}$ and $\underline{1}$ are column vectors with all of their elements being 0 and 1, respectively.

The vector x^* is a *complex balanced equilibrium point* of the closed-loop system if and only if $A_k \cdot \psi(x^*) = \underline{0}$. This is also a *linear constraint*

$$A_k \cdot q^* = \underline{0} \tag{5.41}$$

where $q^* = \psi(x^*)$ is a priori known.

Finally, *by minimizing the objective function* (5.39), *the feedback gain K can be computed (if it exists) in an LP framework* using the linear constraints (5.40), (5.41).

With the resulting feedback gain K, the point x^* will be an equilibrium point of the closed-loop system, and x^* will be locally asymptotically stable in the region $\mathcal{S} = (x^* + S) \cap \overline{\mathbb{R}}_+^n$, where S is the stoichiometric subspace of the closed-loop system. We remark that the stability is proven to be global if the closed-loop system consists of one linkage class [41].

5.2.3 Feedback Computation in the Weakly Reversible Closed Loop With Zero Deficiency Case

In this case, the only computational goal is weak reversibility and zero deficiency, but the equilibrium points are *arbitrary*, that is, they are not specified in advance. We recall that if a kinetic system is weakly reversible and has zero deficiency, then its equilibrium points are complex balanced and they are asymptotically stable within the appropriate stoichiometric compatibility classes. Therefore, the aim of the feedback is to transform the open-loop system into a weakly reversible kinetic system with zero deficiency.

A Kirchhoff matrix A_k is weakly reversible if and only if there exist a positive vector $p \in \mathbb{R}_{>0}^n$ in its kernel. This condition is nonlinear because the vector p is unknown, too. Therefore, we use the solution proposed in [29] as follows. In order to construct linear constraints, we are going to transform the decision variables, the elements of the matrices K and A_k by a diagonal matrix diag$\{p\}$ (a diagonal matrix with the vector elements $p_i > 0$ in its diagonal) where $p \in \overline{\mathbb{R}}_+^n$ to form $A'_k = A_k \cdot \text{diag}\{p\}$ and $K' = K \cdot \text{diag}\{p\}$. Then A'_k has same structure as A_k and $A_k \cdot p = \underline{0}$ if and only if $A'_k \cdot \underline{1} = \underline{0}$. Then the resulting *linear constraints of dynamical equivalence* are

$$
\begin{cases}
M_p \cdot \text{diag}(p) + B \cdot K' = Y \cdot A'_k \\
\underline{1}^T \cdot A'_k = \underline{0}^T \\
A'_{k\,ij} \geq 0 \quad i,j = 1,\ldots,m, \ i \neq j \\
p > \underline{0}
\end{cases}
\tag{5.42}
$$

where the elements of K' and A'_k are the continuous decision variables.

The *linear constraint of weak reversibility* is

$$
A'_k \cdot \underline{1} = \underline{0}
\tag{5.43}
$$

thanks to the suitable transformation of A_k to A'_k.

The MILP constraints of zero deficiency are taken from [54] in the form of

$$
\begin{cases}
\tilde{y}^{(\ell)} = \eta^{(\ell)} + Y^T \cdot \alpha^{(\ell)}, \quad \ell = 1,\ldots,m - \text{rank}(Y) \\
A'_{k\,ij} \leq U_1 \cdot \Theta_{ij}, \quad i,j = 1,\ldots,m \\
|\eta_i^{(\ell)} - \eta_j^{(\ell)}| \leq 2 \cdot U_2(1-\Theta_{ij}), \quad i,j = 1,\ldots,m, \ \ell = 1,\ldots,m-\text{rank}(Y)
\end{cases}
\tag{5.44}
$$

The constant vectors $\tilde{y}^{(\ell)} \in \mathbb{R}^m$ for $\ell = 1,\ldots,m - \text{rank}(Y)$ span the kernel of the matrix Y. The bounds U_1 and U_2 are positive real constants, and by increasing these bounds the optimization problem will be less conservative. The continuous decision variables are $\alpha^{(\ell)} \in \mathbb{R}^n$ and $\eta_i^{(\ell)} \in \mathbb{R}^m$ for $\ell = 1,\ldots,m - \text{rank}(Y)$ and $i = 1,\ldots,m$. In addition, $\Theta_{i,j}$ for $i,j = 1,\ldots,m$ are the binary decision variables.

By minimizing *the objective function* (5.39), *the feedback gain K can be computed (if it exists) in an MILP framework using the linear constraints* (5.42)–(5.44). Note that the resulting equilibrium point can also be computed from the optimization results.

It is important to observe that the design based on achieving a weakly reversible closed loop with zero deficiency has generally less strict constraints—even the equilibrium point cannot be specified—than that of the complex balanced closed-loop case; therefore, it may be feasible when the latter is not. At the same time, the underlying optimization problem is an MILP problem, that is computationally much harder than the LP problem required to solve the feedback design in the complex balanced closed-loop case. However, we can still check the existence of a weakly reversible closed-loop system in polynomial time by checking the feasibility of the linear constraints (5.42), (5.43), since they do not contain integer variables. If no weakly reversible solution exists, then it is unnecessary to run the MILP optimization.

5.3 ROBUSTNESS ISSUES AND ROBUST DESIGN FOR THE STABILIZING CONTROL OF POLYNOMIAL SYSTEMS

In this case, we assume that the coefficient matrix M in the open-loop model (5.35) is not known exactly, but it is an element of the polytopic set

$$\mathcal{M}_p = \left\{ \sum_{i=1}^{L} \alpha_i M_p^{(i)} \mid (\forall i\colon \alpha_i \geq 0) \wedge \sum_{i=1}^{L} \alpha_i = 1 \right\} \qquad (5.45)$$

where $M_p^{(i)} \in \mathbb{R}^{n \times m}$ for $i = 1, \ldots, L$ are the vertex points.

5.3.1 Handling the Parametric Uncertainty of Stabilizing Control of Polynomial Systems in the Complex Balanced Closed-Loop Case

In this section, the feedback computation method described in Section 5.2.2 will be extended by handling parametric uncertainty. The uncertainty is modeled by the polytopic set given in Eq. (5.45).

First, we remark that it is enough to compute a complex balanced realization $(Y, A_k^{(h)})$ with the joint equilibrium point x^* in each vertex $M_p^{(h)}$ ($h = 1, \ldots, L$, where L is the number of vertices of the convex set) with the same feedback gain matrix K (see [27] for more details). This gives the following *constraints*:

$$\begin{cases} M_p^{(h)} + B_p K = Y \cdot A_k^{(h)} \\ \underline{1}^T \cdot A_k^{(h)} = \underline{0}^T \\ A_{k\,ij}^{(h)} \geq 0 \quad i,j = 1, \ldots, m,\ i \neq j \\ A_k^{(h)} \cdot q^* = \underline{0} \end{cases} \qquad (5.46)$$

where $q^* = \psi(x^*)$ is known and $h = 1, \ldots, L$.

Finally, *by minimizing the objective function* (5.39), *the feedback gain K can be computed (if it exists) in an LP framework using the linear constraints* (5.46).

With the resulting feedback gain K, the point x^* will be an equilibrium point of all possible closed-loop systems, and x^* will be locally asymptotically stable in the region $\mathcal{S} = (x^* + S) \cap \overline{\mathbb{R}}_+^n$, where S is the stoichiometric subspace of the closed-loop system.

Example 17 (Robust complex balanced closed-loop design). In this example, the robust design case is considered, when the uncertain coefficient matrix of a polynomial system is characterized as the convex combination constant matrices of appropriate dimensions. Let the open-loop system be given as

$$\dot{x}_p = M_p \underbrace{\begin{bmatrix} x_{p1} x_{p2} \\ x_{p2} x_{p3} \\ x_{p1} \end{bmatrix}}_{\psi_p(x_p)} + \underbrace{\begin{bmatrix} 0 \\ 1 \\ 0 \end{bmatrix}}_{B_p} u_p \tag{5.47}$$

where M_p is the arbitrary convex combination of the following three matrices

$$M_p^{(1)} = \begin{bmatrix} -1 & 1 & 0 \\ 2 & 1 & 2 \\ 1 & -1 & 0 \end{bmatrix}, \quad M_p^{(2)} = \begin{bmatrix} 0 & 0 & 0 \\ 1 & 1 & 3 \\ 0 & 0 & 0 \end{bmatrix}, \quad M_p^{(3)} = \begin{bmatrix} 0 & 1 & -1 \\ 2 & 0 & 3 \\ 0 & -1 & 1 \end{bmatrix}$$

that is, $M_p \in \mathcal{M}_p$.

Let the desired equilibrium point be chosen as $x^* = [1\ 1\ 1]^T$.

We are looking for a feedback law with the gain K that transforms the systems characterized by the matrices $M_p^{(i)}$, $i = 1, 2, 3$ into a complex balanced kinetic system with the given equilibrium point.

By solving the feedback design LP optimization problem using the linear constraints (5.46), we obtain the following feedback:

$$u_p = \begin{bmatrix} -2 & -1 & -2 \end{bmatrix} \psi_p(x_p) \tag{5.48}$$

Fig. 5.1 depicts a complex balanced realization of the closed-loop system in the case $M_p = 0.6 M_p^{(1)} + 0.2 M_p^{(2)} + 0.2 M_p^{(3)}$. The obtained closed-loop system in an inner point of the convex set $\mathcal{M}_p$ has the following stoichiometric subspace:

$$S = \text{span} \left(\begin{bmatrix} -1 \\ 0 \\ 1 \end{bmatrix}, \begin{bmatrix} 0 \\ -1 \\ 0 \end{bmatrix}, \begin{bmatrix} 1 \\ 0 \\ -1 \end{bmatrix}, \begin{bmatrix} 1 \\ -1 \\ -1 \end{bmatrix}, \begin{bmatrix} 0 \\ 1 \\ 0 \end{bmatrix}, \begin{bmatrix} -1 \\ 1 \\ 1 \end{bmatrix} \right) \tag{5.49}$$

the dimension of which is three so it spans the state space.

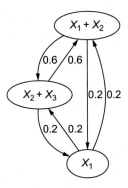

Fig. 5.1 Complex balanced realization of the closed-loop system in the case $M_p = 0.6M_p^{(1)} + 0.2M_p^{(2)} + 0.2M_p^{(3)}$.

Therefore, the equilibrium point x^* will be asymptotically stable in the region $\mathcal{S} = (x^* + S) \cap \overline{\mathbb{R}}_+^n$. Therefore, if the initial value is chosen from the set $\mathcal{S}$, then the corresponding solution will converge to the desired equilibrium point x^*.

Fig. 5.2 shows the time domain behavior of the perturbed system.

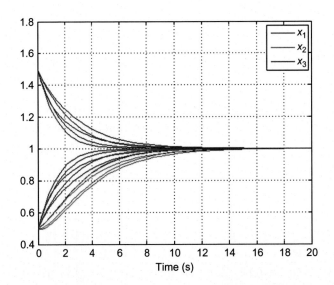

Fig. 5.2 Time domain simulation of the perturbed closed-loop system with the initial value $x_0 = [0.5\ 0.5\ 1.5]$.

CHAPTER 6

Case Studies

6.1 OPTIMIZATION-BASED STRUCTURAL ANALYSIS AND DESIGN OF REACTION NETWORKS

In this section, the algorithms presented in Section 4.5 will be applied for the structural analysis of kinetic biological models. More technical details on the parallelization of the computations can be found in [64]. In this context, by "workers" we mean processes computing reaction graph structures which can be run in a parallel way. The computations were carried

Analysis and Control of Polynomial Dynamic Models with Biological Applications.
https://doi.org/10.1016/B978-0-12-815495-3.00006-7

out on a workstation with two 2.60 GHz Xeon (E5-2650 v2) processors and with 32 Gb RAM (DDR3 1600 MHz, 0.6 ns). The software was written in Python (ver. 2.7.6). Additionally, Python packages such as pyzmq (ver. 14.7.0), cyLP (0.7.2), Cython (ver. 0.23.4), and CBC (ver. 2.8.5) were used. The linear programs were solved with the CLP solver, which is part of CBC.

6.1.1 Computing All Mathematically Possible Structures of a G_1/S Transition Model in Budding Yeast

This example is taken from [65] and models a switch-like behavior in yeast cell cycle regulations. This paper proposes a dynamic kinetic model of the G_1/S transition that uses the following state variables (denoted by $x_i = [X_i]$ for the concentration $[X_i]$ of the specie X_i):

x_1: [Sic1], x_2: [Sic1P], x_3: [Clb], x_4: [Clb·Sic1], x_5: [Clb·Sic1P], x_6: [Cdc14], x_7: [Sic1P·Cdc14], x_8: [Clb·Sic1P·Cdc14], x_9: [Clb·Sic1·Clb]. The short description of the basic components of the species is the following:

- Sic1: Cyclin-dependent kinase inhibitor
- SicP: Phosphorilated form of Sic1
- Clb: B-type cyclin involved in cell cycle progression
- Cdc14: Protein phosphatase required for several functions (e.g., mitotic exit, rDNA segregation, cytokinesis)

More details on the species and the biological background of the modeled process can be found in the original paper [65].

The dynamically equivalent sparse and dense structures of this model (shown in Fig. 6.1) were computed in [52]. The graph structure contains 19 complexes:

$$C_1 = X_2, C_2 = \emptyset, C_3 = X_1, C_4 = X_3 + X_1, C_5 = X_4,$$
$$C_6 = X_3, C_7 = X_2 + X_3, C_8 = X_5, C_9 = X_3 + X_4,$$
$$C_{10} = X_9, C_{11} = X_3 + X_5, C_{12} = X_2 + X_6, C_{13} = X_7,$$
$$C_{14} = X_1 + X_6, C_{15} = X_5 + X_6, C_{16} = X_8,$$
$$C_{17} = X_4 + X_6$$

From the earlier listing of complexes, the complex composition matrix Y can be written as

$$Y = \begin{bmatrix} 0 & 0 & 1 & 1 & 0 & 0 & 0 & 0 & 0 & 0 & 0 & 0 & 0 & 1 & 0 & 0 & 0 \\ 1 & 0 & 0 & 0 & 0 & 0 & 1 & 0 & 0 & 0 & 0 & 1 & 0 & 0 & 0 & 0 & 0 \\ 0 & 0 & 0 & 1 & 0 & 1 & 1 & 0 & 1 & 0 & 1 & 0 & 0 & 0 & 0 & 0 & 0 \\ 0 & 0 & 0 & 0 & 1 & 0 & 0 & 0 & 1 & 0 & 0 & 0 & 0 & 0 & 0 & 0 & 1 \\ 0 & 0 & 0 & 0 & 0 & 0 & 0 & 1 & 0 & 0 & 1 & 0 & 0 & 0 & 1 & 0 & 0 \\ 0 & 0 & 0 & 0 & 0 & 0 & 0 & 0 & 0 & 0 & 0 & 1 & 0 & 1 & 1 & 0 & 1 \\ 0 & 0 & 0 & 0 & 0 & 0 & 0 & 0 & 0 & 0 & 0 & 0 & 1 & 0 & 0 & 0 & 0 \\ 0 & 0 & 0 & 0 & 0 & 0 & 0 & 0 & 0 & 0 & 0 & 0 & 0 & 0 & 0 & 1 & 0 \\ 0 & 0 & 0 & 0 & 0 & 0 & 0 & 0 & 0 & 1 & 0 & 0 & 0 & 0 & 0 & 0 & 0 \end{bmatrix}$$

$$\tag{6.1}$$

The nonzero off-diagonal elements of A_k are the following:

$$A_{k\,2,1} = k_3,\ A_{k\,2,3} = k_2,\ A_{k\,3,2} = k_1,$$
$$A_{k\,4,5} = k_5,\ A_{k\,5,4} = k_4,\ A_{k\,6,5} = k_6,$$
$$A_{k\,6,8} = k_9,\ A_{k\,7,8} = k_8,\ A_{k\,8,7} = k_7, \tag{6.2}$$
$$A_{k\,9,10} = k_{11},\ A_{k\,10,9} = k_{10},\ A_{k\,11,10} = k_{12},$$
$$A_{k\,12,13} = k_{14},\ A_{k\,13,12} = k_{13},\ A_{k\,14,13} = k_{15},$$
$$A_{k\,15,16} = k_{17},\ A_{k\,16,15} = k_{16},\ A_{k\,17,16} = k_{18}$$

For the computations, we used the same rate coefficients as in [52], namely

$$k = [4.1\ 3.2\ 6.7\ 7.3\ 3.8\ 2.4\ 4.5\ 5.1\ 6.2$$
$$7.7\ 8.6\ 9.5\ 2.4\ 4.9\ 5.8\ 10.2\ 6.3\ 8.5]^T \tag{6.3}$$

The dense realization contains 28 reactions, while the sparse has 18. It should be pointed out that the original realization of the model reported in [65] contains 18 reactions, that is, this realization is sparse, and it is the only sparse realization of the given dynamics.

With 31 workers, the computation of this model took 7.66 s, and we found 729 different realization structures of the kinetic system. Fig. 6.2 shows the number of different reaction graph structures containing a given number of reactions. We can see that the sparse structure is unique in this case.

Given all realizations we can easily calculate the deficiency of each realization (see Table 6.1). The complex C_6 becomes isolated in 81 realizations—these are the ones with deficiency 2. The role of C_6 is then taken up by combinations of reactions originating from complexes C_5 and C_8 in this case. The possible combinations must be the subgraphs of the dense realization (shown in the right panel of Fig. 6.1).

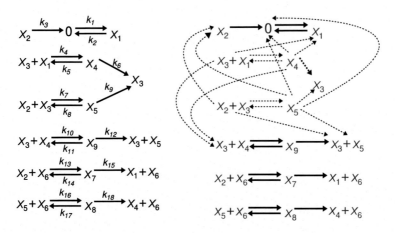

Fig. 6.1 *Left: The originally published model of the G_1/S transition model from [65], which is also a sparse realization. Right: The dynamical equivalent dense realization of the same model.*

(This figure is taken from G. Szederkényi, J.R. Banga, A.A. Alonso, Inference of complex biological networks: distinguishability issues and optimization-based solutions, BMC Syst. Biol. 5 (2011) 177, https://doi.org/10.1186/1752-0509-5-177.)

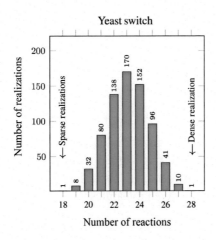

Fig. 6.2 *Distribution of the number of different reaction graph structures over the number of reactions for the G_1/S transition model.*

Table 6.1 Number of Graph Structures With Different Deficiencies for the G_1/S Transition Model	
Deficiency	**Number of Realizations**
2	81
5	19
6	629

6.1.2 Analysis of a Five-Node-Repressilator With Auto Activation

This model was reported in [66], and it was adapted and investigated in details in [52, 64]. The model contains five genes with auto-activation, and each gene represses another gene. The layout, connection pattern of genes, and the corresponding reactions are shown in Fig. 6.3.

Similar to [66], we assume the following:

- cooperative regulator binding,
- genes are present in constant amounts,
- transcription and translation are modeled by single-step kinetics, and finally,
- proteins are degraded by first-order reactions.

We note that complex dynamic phenomena such as multiple steady states or oscillations have been shown with a wide range of parameters in similar systems, especially in the case when the number of genes is odd [66]. We also assume that there is some protein production (leakage) when both the activator and the repressor are bound to the genes (although this assumption does not affect the main results of the forthcoming analysis).

Using the assumptions listed earlier, the CRN describing the system is the following:

$$G_i + P_i \underset{k_{i,2}}{\overset{k_{i,1}}{\rightleftarrows}} G_i^A \quad \text{(auto-activation 1)} \tag{6.4}$$

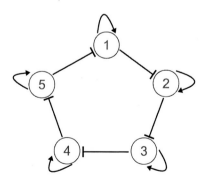

Fig. 6.3 Left: Layout and connection pattern of the 5-node-repressilator. Each gene labeled with numbers 1 to 5 consists of 11 reactions responsible for gene expression, repression, and protein degradation as listed on the right panel.
(This figure is taken from G. Szederkényi, J.R. Banga, A.A. Alonso, Inference of complex biological networks: distinguishability issues and optimization-based solutions, BMC Syst. Biol. 5 (2011) 177, https://doi.org/10.1186/1752-0509-5-177.)

$$G_i^A \xrightarrow{k_{i,3}} G_i^A + P_i \text{ (protein production 1)} \tag{6.5}$$

$$G_i + P_j \underset{k_{i,5}}{\overset{k_{i,4}}{\rightleftharpoons}} G_i^R \text{ (repression 1)} \tag{6.6}$$

$$G_i^R + P_i \underset{k_{i,7}}{\overset{k_{i,6}}{\rightleftharpoons}} G_i^{AR} \text{ (auto-activation 2)} \tag{6.7}$$

$$G_i^A + P_j \underset{k_{i,9}}{\overset{k_{i,8}}{\rightleftharpoons}} G_i^{AR} \text{ (repression 2)} \tag{6.8}$$

$$G_i^{AR} \xrightarrow{k_{i,10}} G_i^{AR} + P_i \text{ (protein production 2)} \tag{6.9}$$

$$P_i \xrightarrow{k_{i,11}} 0 \text{ (protein degradation)} \tag{6.10}$$

for the index pairs $(i,j) \in \{(1,5),(2,1),(3,2),(4,3),(5,4)\}$. In Eqs. (6.4)–(6.10), G_i and P_i represent the ith gene and protein, respectively. For the genes, superscripts A and R refer to activated and repressed states, respectively. Let us denote with $r_{i,k}$ the reaction with rate coefficient $k_{i,k}$ in Eqs. (6.4)–(6.10).

Two cases with different sets of randomly selected rate coefficients were studied, and the structures of the obtained results were the same. The numerical details can be found on the third sheet of Additional File 1 of [52].

The total number of reactions for the repressilator model is 55; that is equal to the number of reactions in the sparse realization. The dense realization contains 70 reactions, which means that there are a maximum of 15 more mathematically possible reactions while maintaining exactly the same dynamics as the original biological model. These additional reactions are the following:

$$G_i^{AR} \to G_i^R, \quad P_i + G_i^R \to G_i^R$$
$$P_i + G_i^R \to P_i + G_i^{AR}, \quad \text{for } i = 1,\dots,5 \tag{6.11}$$

The number of core reactions in the model are 45. The set of noncore reactions (that, in principle, can be substituted by other reactions) is given by

$$G_i^{AR} \leftrightarrows G_i^R + P_i, \quad i = 1, \ldots, 5 \tag{6.12}$$

In particular, it is easy to show (see also Additional File 1 of [52]) that reactions $G_i^{AR} \rightarrow G_i^R + P_i$ and $G_i^{AR} \rightarrow G_i^R$ are always indistinguishable. Similarly, the reaction $G_i^R + P_i \rightarrow G_i^{AR}$ can be substituted with the combination of reactions $G_i^R + P_i \rightarrow G_i^R$ and $G_i^R + P_i \rightarrow G_i^{AR} + P_i$.

As in the previous example in Section 6.1.1, the dense-sparse gap indicates the possibility of different realizations. After computing all realizations we ended up with 65,071 structurally different realizations. The computation took 554 min; this is due to the size of the linear program, which has 5202 decision variables.

Finally, we can again investigate the distribution of the number of reactions among the distinct realizations (Fig. 6.4). It is worth remarking that the number of sparse structures leading to the same dynamics is 46 in this case. This shows that a sparsity assumption alone is generally not sufficient for ensuring structural uniqueness.

6.1.3 Different Realizations of an Oscillating Rational System

Here, we illustrate the notions of Section 2.3.4 related to the graph structure of rational kinetic models. The example is taken from [26], and it is a modified version of the mass action system corresponding to Example A1 in [67].

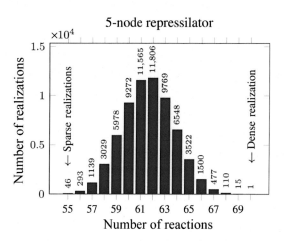

Fig. 6.4 Distribution of the number of different reaction graph structures over the number of reactions for the five-node repressilator model.

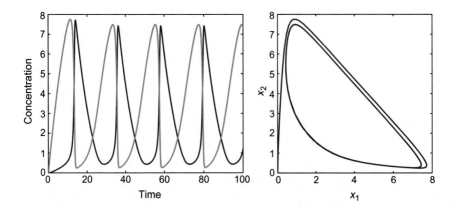

Fig. 6.5 *Solution of the dynamic Eq. (6.13) in the time domain and in the phase space.*

$$\frac{dx_1}{dt} = 0.05x_2 + 0.1x_1^2x_2 - \frac{0.2x_1^2}{1 + 2x_1} - \frac{x_1}{1 + \frac{1}{2}x_1}$$

$$\frac{dx_2}{dt} = 1 - 0.05x_2 - 0.1x_1^2x_2 + \frac{0.1x_1^2}{1 + 2x_1}$$

(6.13)

The earlier set of dynamic equations gives rise to a stable limit cycle solution, as depicted in Fig. 6.5 for zero initial condition.

A biochemical reaction network realization, which corresponds to the previous dynamic equations, can be constructed in the following way. First, we assume the complex set

$$\mathcal{C} = \{0, X_1, X_2, 2X_1, 2X_1 + X_2, 3X_1\}$$

as in [67], where 0 denotes the so-called zero complex (see, e.g., [16]). Then, for each complex $C_1, \ldots, C_6$, a kinetic rate function is derived as

$$g_{11} = 1, \; g_{21} = \frac{x_1}{1 + x_1/2}, \; g_{31} = x_2, \; g_{41} = \frac{x_1^2}{1 + 2x_1}, \; g_{51} = x_1^2x_2, \; g_{61} = x_1^3$$

respectively. This means mass action type rate laws for complexes C_1, C_3, and C_5, and Michaelis-Menten-type reaction kinetics for the complexes C_2 and C_4.

6.1.3.1 Dynamically Equivalent Representation
The matrix representation of the kinetic system in Eq. (6.13) can be written as

$$Y = \begin{bmatrix} 0 & 1 & 0 & 2 & 2 & 3 \\ 0 & 0 & 1 & 0 & 1 & 0 \end{bmatrix}, \quad A_k = \begin{bmatrix} -k_1 & k_2 & 0 & 0 & 0 & 0 \\ 0 & -k_2 & k_3 & 0 & 0 & 0 \\ k_1 & 0 & -k_3 & k_4 & 0 & 0 \\ 0 & 0 & 0 & -k_4 & 0 & 0 \\ 0 & 0 & 0 & 0 & -k_5 & 0 \\ 0 & 0 & 0 & 0 & k_5 & 0 \end{bmatrix},$$

$$P(x) = \begin{bmatrix} 1 & 0 & 0 & 0 & 0 & 0 \\ 0 & \frac{1}{1+x_1/2} & 0 & 0 & 0 & 0 \\ 0 & 0 & 1 & 0 & 0 & 0 \\ 0 & 0 & 0 & \frac{1}{1+2x_1} & 0 & 0 \\ 0 & 0 & 0 & 0 & 1 & 0 \\ 0 & 0 & 0 & 0 & 0 & 1 \end{bmatrix}, \quad \Psi(x) = \begin{bmatrix} 1 \\ x_1 \\ x_2 \\ x_1^2 \\ x_1^2 x_2 \\ x_1^3 \end{bmatrix}$$

where the parameter values are $k_1 = 1$, $k_2 = 1$, $k_3 = 0.05$, $k_4 = 0.1$, and $k_5 = 0.1$.

This is a sparse realization of the dynamical equations with five reactions (i.e., it contains the minimum number of reactions), and its reaction graph is shown in Fig. 6.6.

A dynamically equivalent dense realization of the earlier can be found with 15 reactions using the optimization framework in [25]. This reaction graph is depicted in Fig. 6.7. Its generalized Kirchhoff matrix is given by

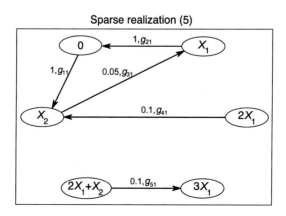

Fig. 6.6 Sparse realization.

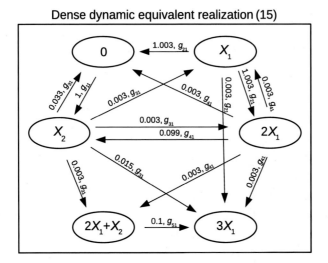

Fig. 6.7 Dense realization.

$$A_k^{\text{dense}} = \begin{pmatrix} -1.00 & 1.003 & 0.033 & 10^{-3} & 0 & 0 \\ 0 & -1.005 & 10^{-3} & 10^{-3} & 0 & 0 \\ 1.00 & 0 & -0.051 & 0.099 & 0 & 0 \\ 0 & 10^{-3} & 10^{-3} & -0.103 & 0 & 0 \\ 0 & 0 & 10^{-3} & 10^{-3} & -0.10 & 0 \\ 0 & 10^{-3} & 0.015 & 10^{-3} & 0.10 & 0 \end{pmatrix}$$

It was shown in [25] that the dense dynamically equivalent realization has a unique structure, and further, that any dynamically equivalent realization contains a *subset* of the reactions of the dense realization; that is, the reactions which are not included in the dense reaction cannot appear in any other dynamically equivalent realization.

6.1.4 Linearly Conjugate Structures of the Model

Now, we compute a dense linearly conjugate realization of the dynamic system (6.13). First, we solve the optimization problem described in Section 4.5.5 with the following computational parameters:

- threshold value $\epsilon = 10^{-6}$
- bound constraints for A_b: $a_{ij}^{\text{lb}} = 0$ and $a_{ij}^{\text{ub}} = 100$ for $i = 1, \ldots, m$ and $j = 1, \ldots, \kappa$,
- bound constraints for $\tilde{c}$: $\tilde{c}_i^{\text{lb}} = 0.01$ and $\tilde{c}_i^{\text{ub}} = 1000$ for $i = 1, \ldots, m$ and $j = 1, \ldots, \kappa$.

The solution of the optimization resulted in the following Kirchhoff and scaling matrices

$$
A_b = \begin{pmatrix}
-0.24 & 0.163 & 9.0 \times 10^{-3} & 10^{-3} & 10^{-3} & 0 \\
0 & -0.165 & 10^{-3} & 10^{-3} & 10^{-3} & 0 \\
0.24 & 0 & -0.013 & 0.015 & 10^{-3} & 0 \\
0 & 10^{-3} & 10^{-3} & -0.027 & 10^{-3} & 0 \\
0 & 0 & 10^{-3} & 9.000 \times 10^{-3} & -0.025 & 0 \\
0 & 10^{-3} & 10^{-3} & 10^{-3} & 0.021 & 0
\end{pmatrix}
$$

$$
T(x)^{-1} = \begin{pmatrix} 0.16 & 0 \\ 0 & 0.24 \end{pmatrix}
$$

Then, following the procedure described in Section 4.5.3, first we obtain T and matrix $\bar{\Phi}_T$ by Eq. (3.19), as

$$
T = \begin{pmatrix} 6.25 & 0 \\ 0 & 4.1667 \end{pmatrix},
$$

$$
\bar{\Phi}_T = \mathrm{diag}\left(\left[1,\ 0.16,\ 0.24,\ 0.0256,\ 6.144 \times 10^{-3},\ 4.096 \times 10^{-3}\right]^T\right)
$$

Finally, Eqs. (3.17), (3.18) give the dense generalized Kirchhoff matrix of the linearly conjugate network

$$
\bar{A}_k^{\mathrm{dense}} = \begin{pmatrix}
-0.2400 & 1.019 & 0.03750 & 0.03906 & 0.1628 & 0 \\
0 & -1.031 & 4.167 \times 10^{-3} & 0.03906 & 0.1628 & 0 \\
0.2400 & 0 & -0.05417 & 0.5859 & 0.1628 & 0 \\
0 & 6.250 \times 10^{-3} & 4.167 \times 10^{-3} & -1.055 & 0.1628 & 0 \\
0 & 0 & 4.167 \times 10^{-3} & 0.3516 & -4.069 & 0 \\
0 & 6.250 \times 10^{-3} & 4.167 \times 10^{-3} & 0.03906 & 3.418 & 0
\end{pmatrix}
$$

and its transformed kinetic weighting function

$$
\bar{P}(x) = \begin{pmatrix}
1 & 0 & 0 & 0 & 0 & 0 \\
0 & \frac{1}{1+6.25 \cdot 0.5x_1} & 0 & 0 & 0 & 0 \\
0 & 0 & 1 & 0 & 0 & 0 \\
0 & 0 & 0 & \frac{1}{1+6.25 \cdot 2x_1} & 0 & 0 \\
0 & 0 & 0 & 0 & 1 & 0 \\
0 & 0 & 0 & 0 & 0 & 1
\end{pmatrix}
$$

The reaction graph of the resulted linearly conjugate network $(Y, \bar{A}_k^{\mathrm{dense}}, \bar{P})$ is depicted in Fig. 6.8. It can be seen that there are four edges, that is,

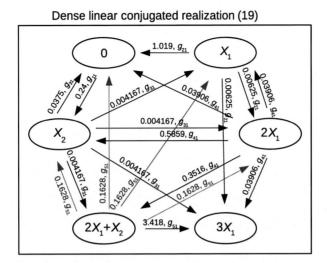

Fig. 6.8 Linearly conjugate dense realization of the oscillating dynamics.

four reactions, highlighted in red color, which do not appear in the dense dynamically equivalent realization (see Fig. 6.7).

For the comparison of the dynamics of the original and the linearly conjugate network, the solutions corresponding to the zero initial condition are depicted in the time domain and in the phase space; see Fig. 6.9, where we can see the effect of scaling and the qualitative similarities with the original solutions in Fig. 6.5.

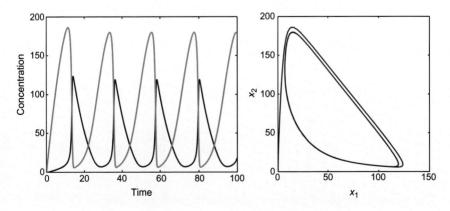

Fig. 6.9 Solution of the linearly conjugate dynamic equations in the time domain and in the phase space.

6.2 COMPUTATIONAL DISTINGUISHABILITY ANALYSIS OF AN UNCERTAIN KINETIC MODEL

In this section, the possible reaction graph structures of an uncertain G-protein cycle model taken from the literature will be examined.

6.2.1 Biological Background

The G-protein (guanine nucleotide-binding protein) cycle has a key role in several intracellular signaling transduction pathways. The G-protein located on the intracellular surface of the cell membrane is activated by the binding of specific ligand molecules to the G-protein-coupled receptor of the extracellular membranes' surface. The activated G-protein dissociates to different subunits which take part in intracellular signaling pathways. After the termination of the signaling mechanisms, the subunits become inactive and bind into each other [68].

We examined the structural properties of the yeast G-protein cycle using the model published in [69]. The model involves a so-called heterotrimeric G-protein containing three different subunits. In response to the extracellular ligand binding, the protein dissociates to G-α and G-$\beta\gamma$ subunits, where the active and inactive forms of the G-α subunit can also be distinguished.

The reaction network model involves the following species: R and L represent the receptor and the corresponding ligand, respectively; RL refers to the ligand-bound receptor; G is the G-protein located on the intracellular membrane surface; G_a and G_d denote the active and the inactive forms of the G-α subunit; and G_{bg} is the G-$\beta\gamma$ subunit.

6.2.2 Kinetic Representation

The model can be characterized as a chemical reaction network (Y, A_k), where the structures of the complexes and the reactions are defined by the complex composition matrix $Y \in \mathbb{R}^{7 \times 10}$ and the Kirchhoff matrix $A_k \in \mathbb{R}^{10 \times 10}$ are as follows:

$$Y = \begin{bmatrix} 1 & 0 & 1 & 0 & 0 & 0 & 0 & 0 & 0 & 0 \\ 0 & 0 & 1 & 0 & 0 & 0 & 0 & 0 & 0 & 0 \\ 0 & 1 & 0 & 0 & 0 & 0 & 1 & 0 & 0 & 0 \\ 0 & 0 & 0 & 1 & 0 & 0 & 1 & 0 & 0 & 0 \\ 0 & 0 & 0 & 0 & 1 & 0 & 0 & 1 & 0 & 0 \\ 0 & 0 & 0 & 0 & 0 & 0 & 0 & 1 & 1 & 0 \\ 0 & 0 & 0 & 0 & 0 & 1 & 0 & 0 & 1 & 0 \end{bmatrix}$$

$$A_k = \begin{bmatrix} -0.4 & 0 & 0 & 0 & 0 & 0 & 0 & 0 & 0 & 4000 \\ 0 & -14 & 0.322 & 0 & 0 & 0 & 0 & 0 & 0 & 0 \\ 0 & 10 & -0.322 & 0 & 0 & 0 & 0 & 0 & 0 & 0 \\ 0 & 0 & 0 & 0 & 0 & 0 & 0 & 0 & 1000 & 0 \\ 0 & 0 & 0 & 0 & -11{,}000 & 0 & 0 & 0 & 0 & 0 \\ 0 & 0 & 0 & 0 & 11{,}000 & 0 & 0 & 0 & 0 & 0 \\ 0 & 0 & 0 & 0 & 0 & 0 & -0.01 & 0 & 0 & 0 \\ 0 & 0 & 0 & 0 & 0 & 0 & 0.01 & 0 & 0 & 0 \\ 0 & 0 & 0 & 0 & 0 & 0 & 0 & 0 & -1000 & 0 \\ 0.4 & 4 & 0 & 0 & 0 & 0 & 0 & 0 & 0 & -4000 \end{bmatrix}$$

The kinetic system that is realized by the model is $\dot{x} = M_N \cdot \psi^Y = Y \cdot A_k \cdot \psi^Y$ (i.e., $M_N = Y \cdot A_k \in \mathbb{R}^{7 \times 10}$), where the notation M_N refers to the nominal value of the coefficient matrix M

The reaction graph structure of the G-protein model can be seen in Fig. 6.10 with the indication of the linkage classes. (The linkage classes are the undirected connected components of the reaction graph.) The computation of all possible reaction graph structures and the solution of the linear equations shows that the heterotrimeric G-protein cycle with the given parametrization is not just structurally, but also parametrically unique. Thus, the prescribed dynamics without uncertainty cannot be realized by any other set of reactions or different reaction rate coefficients using the given set of complexes.

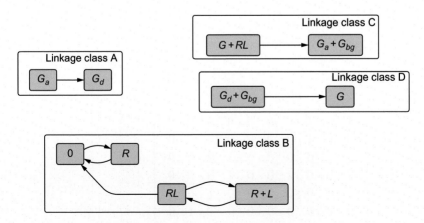

Fig. 6.10 Reaction graph structure of the heterotrimeric G-protein cycle.

6.2.3 Analysis of the Uncertain Model
As is common in biological applications, the model parameters can only be determined with large uncertainties from the measurements. This has a major impact on the possible kinetic structures and their properties.

6.2.3.1 Uncertainty Representation
The uncertainty model used here represents a special case in the class of uncertain kinetic systems, since the possible values of every coefficient of the kinetic system are determined by independent upper and lower bounds that are defined as relative distances. The uncertainty representation follows the method described earlier in Section 4.5.6 using polyhedrons (see in Section B.4.5).

Let us represent the (i,j)th entry the uncertain coefficient matrix M by the coordinate $\tilde{M}_l$ of the point $\tilde{M} \in \mathbb{R}^{n \cdot m}$. Moreover, let the relative distances of the upper and lower bounds of $\tilde{M}_l$ be given by the real constants γ_l and ρ_l from the interval $[0,1]$, respectively. Then the equations defining the polyhedron $\mathcal{P} \subset \mathbb{R}^{n \cdot m}$ of the uncertain parameters can be written in terms of the coordinates $\tilde{M}_l$ as

$$\tilde{M}^\top \cdot e_l^{n \cdot m} \leq (1 + \gamma_l) \cdot [M_N]_{ij}, \quad l = 1, \ldots, n \cdot m \qquad (6.14)$$

$$\tilde{M}^\top \cdot (-e_l^{n \cdot m}) \leq (\rho_l - 1) \cdot [M_N]_{ij}, \quad l = 1, \ldots, n \cdot m \qquad (6.15)$$

where $e_l^{n \cdot m}$ denotes the lth standard basis vector in $\mathbb{R}^{n \cdot m}$. The resulting uncertain kinetic system is denoted by $[\mathcal{P}, L, Y]$. If no additional linear constraints are considered, then $L = \emptyset$ (see Section 4.5.6).

6.2.3.2 Results for the Uncertain Model Without Additional Constraints
In this case, the effect of the growing uncertainty is investigated on the possible kinetic structures.

10% Uncertainty for All Parameters
First, we examined the uncertain model $[\mathcal{P}_1, \emptyset, Y]$, where the uncertainty coefficients γ_l and ρ_l for all $l \in \{1, \ldots, 70\}$ are 0.1 and there are no additional linear constraints in the model. By computing all possible reaction graph structures and the set of core reactions of this uncertain kinetic system, we obtained that all the reactions in the original G-protein cycle are core reactions. Moreover, in the dense realization there are 10 further reactions, and these can be present in the realizations independently of each other. Consequently, the total number of different graph structures is $2^{10} = 1024$.

Fig. 6.11 shows the number of possible reaction graph structures with different number of reactions. The dense realization (that is always unique) is shown in Fig. 6.12 for this case.

20% Uncertainty for All Parameters
If we increase the relative uncertainty to 0.2, we obtain the uncertain kinetic system $[\mathcal{P}_2, \emptyset, Y]$ with $\gamma_l = \rho_l = 0.2$ for $i = 1, \ldots, 70$. In this case, the

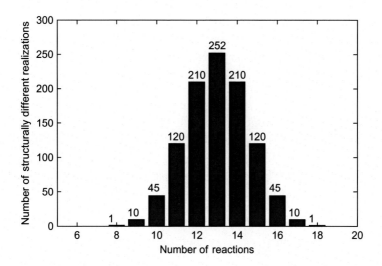

Fig. 6.11 The number of structurally different realizations of the uncertain kinetic system $[\mathcal{P}_1, \emptyset, Y]$ with different number of reactions.

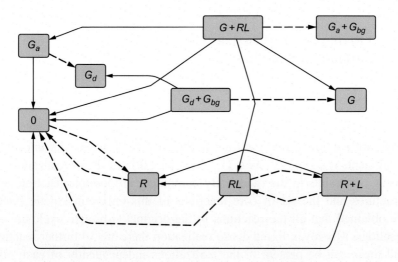

Fig. 6.12 Reaction graph structure of the dense realization of the uncertain kinetic system $[\mathcal{P}_1, \emptyset, Y]$. The core reactions are drawn with dashed lines.

reaction $RL \to 0$ is no longer a core reaction, and it can also be added or removed independently of all other reactions (which remain independent of each other). Therefore, the number of possible structures becomes $2^{11} = 2048$.

6.2.3.3 Constrained Uncertain Model

We have also examined the possible structures in the case of constrained uncertain models. The set of constraints L_1 was chosen to exclude every reaction between different linkage classes.

It can be seen in Fig. 6.13 that the dense realization of the uncertain kinetic system $[\mathcal{P}_1, L_1, Y]$ with the uncertainty coefficients γ_l and ρ_l for all $l \in \{1, \ldots, 70\}$ equal to 0.1 has three reactions that are exactly the ones that are present in the dense realization of $[\mathcal{P}_1, \emptyset, Y]$ and do not connect different linkage classes. These reactions are independent of each other; therefore, the number of structurally different realizations is $2^3 = 8$ in the case of the uncertain kinetic system $[\mathcal{P}_1, L_1, Y]$ and $2^4 = 16$ for the model $[\mathcal{P}_2, L_1, Y]$. The sets of core reactions are the same as in the case of the unconstrained model for both degrees of uncertainty.

The independence of noncore reactions is a special property of the studied uncertain model. As a consequence of this and the superstructure property of the dense realization, the dense realization of the constrained model will

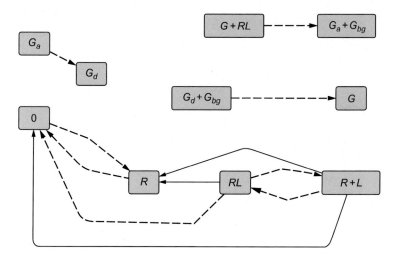

Fig. 6.13 Reaction graph structure of the dense realization of the uncertain kinetic system $[\mathcal{P}_1, L_1, Y]$. The core reactions are drawn with dashed lines.

contain each reaction of the unconstrained model that is not excluded by the constraints.

We emphasize that the dense realizations in the earlier example contain all mathematically possible reactions that can be compatible with the studied uncertain models. If, using prior knowledge, the biologically nonplausible reactions are excluded, and/or certain relations between model parameters are ensured via linear constraints, then the described methodology is still suitable to check the structural uniqueness of the resulting uncertain kinetic model.

6.2.3.4 Analysis of the Number of Realizations for Different Degrees of Uncertainty

It is also worth examining the number of possible reaction graph structures as we increase the uncertainty interval. The largest relative uncertainty examined was defined by $\gamma_l = \rho_l = 0.9$. For this uncertainty, the dense realization is shown in Fig. 6.14. It can be seen that the reactions and core reactions in this case are exactly the same as those were described in Section 6.2.3.2. In other words, the degree of structural uncertainty does not increase between 20% and 90% relative parameter uncertainty.

This fact is further supported by Fig. 6.15 which shows the number of possible reaction graph structures and core reactions for different degrees of relative parameter uncertainty. Because the more than 2000 possible structures for relative uncertainties larger than or equal to 0.2 shown in

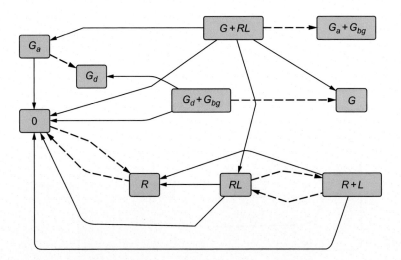

Fig. 6.14 *Reaction graph structure of the dense realization of the unconstrained G-protein model for ±0.9 relative uncertainty. The core reactions are drawn with dashed lines.*

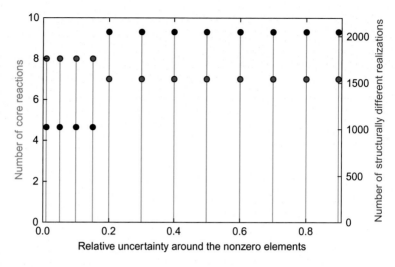

Fig. 6.15 The number of structurally different realizations and number of core reactions as a function of relative uncertainty around the nonzero entries of M.

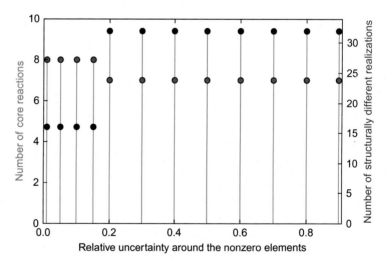

Fig. 6.16 The number of structurally different realizations and number of core reactions as a function of relative uncertainty around the nonzero entries of M. The reactions between complexes belonging to different linkage classes are not allowed.

Fig. 6.15 are quite large from a practical point of view, we also examine the number of different structures for the constrained model described in Section 6.2.3.3. The results are shown in Fig. 6.16. We can see that appropriately chosen structural and/or parametric constraints can efficiently reduce the number of reaction graphs corresponding to an uncertain kinetic model.

6.3 STABILITY ANALYSIS AND STABILIZING CONTROL OF FERMENTATION PROCESSES IN QP FORM

In the following, some simple process system examples are proposed for the stability analysis and the stabilizing controller design problem for quasipolynomial (QP) and Lotka-Volterra (LV) systems discussed so far. The investigated systems are simple, continuously stirred tank reactor (CSTR) examples with different types of fermentation processes.

6.3.1 Zero Dynamics of the Simple Fermentation Process

The next case study indicates that a fortunate choice of a QP-type feedback can simplify the dynamics of a closed-loop system in such a way that the number of quasimonomials may drastically decrease.

Let us consider a single input single output (SISO) input-affine QP model in the form of Eq. (5.22) with $r = 1$, that is

$$\frac{dx_i}{dt} = x_i \left(\lambda_{0_i} + \sum_{j=1}^{m} \mathcal{A}_{0_{i,j}} \prod_{k=1}^{n} x_k^{B_{j,k}} \right)$$

$$+ x_i \left(\lambda_{1_i} + \sum_{j=1}^{m} \mathcal{A}_{1_{i,j}} \prod_{k=1}^{n} x_k^{B_{j,k}} \right) \cdot u, i = 1, \ldots, n \qquad (6.16)$$

where the vector of state variables x and the input variable u are defined on the nonnegative orthant, and $p_j = \prod_{k=1}^{n} x_k^{B_{j,k}}$ for $j = 1, \ldots, m$ are the quasimonomials. Let us assume the simplest output $y = x_i - w^*$ for some i and $w^* > 0$, that is, we want to keep the system's output being equal to a state variable at a positive constant value. Moreover, let us assume that the relative degree of the system equals one and $g_{i1}(x) = g_i(x) = \prod_{j=1}^{n} x_j^{\gamma_{ji}}$, that is, the input function is of quasimonomial type (see Chapter 5 and Eq. (6.16) above). Then the output zeroing input is given in the form

$$u(t) = -\frac{L_f h(x)}{L_g h(x)} = -\frac{f_i(x)}{\prod_{j=1}^{n} x_j^{\gamma_{ji}}} \qquad (6.17)$$

It is seen that the output zeroing input earlier is in QP form if $f_i(x)$ is in QP form.

In order to obtain the zero dynamics (see Appendix B.4.4 and [70]), one has to substitute the input Eq. (6.17) to the state Eq. (6.16) to obtain an *autonomous* system model. It is easy to compute that the *resulting zero dynamics system model will remain in QP form with an output zeroing input in QP form* [71].

Therefore, the stability analysis of the zero dynamics can be investigated using the methods described in Section 4.1.2. The earlier result can be easily generalized to the case of output functions in quasimonomial form.

In what follows, a slightly different version of the fermentation model (3.48) is examined where the input is the flowrate F. The values of S_F and X_F are the constant steady-state values. The moel parameter values collected in Table 6.2 were used.

The zero dynamics analysis for the fermentation example can be performed, for example, by using the output

$$y = x_2 - x_2^*$$

that is, the centered substrate concentration. The output zeroing input can be easily computed

$$F = \frac{\mu_{\max}x_2^* V}{Y(S_F - x_2^*)}x_1\eta \tag{6.18}$$

If the previous equations are substituted into the QP form (3.48), one gets the following zero dynamics

$$\dot{x}_1 = x_1\left(\frac{\mu_{max}x_2^*}{K_2x_2^{*2} + x_2^* + K_1} - \frac{x_2^*\mu_{max}}{Y(S_F - x_2^*)(K_2x_2^{*2} + x_2^* + K_1)}x_1\right) \tag{6.19}$$

Table 6.2 Variables and Parameters of the Fermenter Model With Nonmonotonous Kinetics (3.48)

x_1	Biomass concentration		g/L
x_2	Substrate concentration		g/L
S_F	Substrate feed concentration		g/L
X_F	Biomass feed concentration		g/L
F	Inlet feed flow-rate	3.2089	L/h
V	Volume	4.0000	L
Y	Yield coefficient	0.5000	–
$\mu_{\max}$	Kinetic parameter	1.0000	L/h
K_1	Kinetic parameter	0.0300	g/L
K_2	Kinetic parameter	0.5000	L/g

with QP matrices $\mathcal{A}'$, $\mathcal{B}'$, and λ' being the following ones:

$$\mathcal{A}' = \left[\ -\frac{x_2^*\mu_{\max}}{Y(S_F - x_2^*)(K_2 x_2^{*2} + x_2^* + K_1)} \ \right] = \left[\ -0.1640 \ \right],$$

$$\mathcal{B}' = \left[\ 1 \ \right], \quad \lambda' = \left[\ \frac{\mu_{\max} x_2^*}{K_2 x_2^{*2} + x_2^* + K_1} \ \right] = \left[\ 0.8022 \ \right]$$

Hence, the only monomial of the zero dynamics is

$$x_1 = X$$

where X is the biomass concentration.

Note that *the number of quasimonomials has been drastically reduced.*

In order to study the local stability of the zero dynamics, we first computed the eigenvalue (i.e., the value) of the Jacobian of the zero dynamics at the equilibrium point x_1^* that is

$$- 0.8022$$

Thereafter, the feasibility of the LMI (4.6) was investigated using the LMI Toolbox in Matlab for global stability analysis. The result of the LMI is the following Lyapunov function parameter matrix:

$$C = \left[\ 2.7642 \ \right]$$

Therefore, the global stability of the zero dynamics is proved through the QP description. This result is in good agreement with [72], where the stability of the zero dynamics was proved through nonlinear coordinates-transformations.

6.3.2 Partially Actuated Fermentation Example in QP Form
The system of this example is a simpler variant of the fermentation process (3.48) of Section 3.4 with S_F being the manipulable input in the following form:

$$\begin{aligned}
\dot{x}_1 &= \mu_{\max} x_1 x_2 - \frac{F}{V} x_1 \\
\dot{x}_2 &= -\frac{\mu_{\max}}{Y} x_1 x_2 + \frac{F}{V}(S_F - x_2)
\end{aligned} \tag{6.20}$$

The parameter values can be seen in Table 6.3.

The QP form of the model is:

$$\begin{aligned}
\dot{x}_1 &= x_1 (x_2 - 2) \\
\dot{x}_2 &= x_2 \left(-x_1 + 2x_2^{-1} S_F - 2 \right)
\end{aligned} \tag{6.21}$$

Table 6.3 Variables and Parameters of the Fermenter Model (6.20)

x_1	Biomass concentration		g/L
x_2	Substrate concentration		g/L
F	Inlet feed flow-rate	2	L/h
V	Volume	1	L
S_F	Substrate feed concentration		g/L
Y	Yield coefficient	1	–
μ_{max}	Kinetic parameter	1	L/h

For a fixed value of the substrate concentration $S_F^* = 1$, the system has an asymptotically stable washout-type equilibrium point

$$\begin{bmatrix} x_1^* \\ x_2^* \\ S_F^* \end{bmatrix} = \begin{bmatrix} 0 \\ 1 \\ 1 \end{bmatrix}$$

The feedback structure was chosen to be

$$S_F = k_1 x_2^2 + \delta_1 x_2$$

The closed-loop system with the earlier structure is

$$\dot{x}_1 = x_1 (x_2 - 2)$$
$$\dot{x}_2 = x_2 (-x_1 + 2k_2 x_2 + 2(\delta_1 - 1)) \tag{6.22}$$

It is apparent that the earlier QP model (6.22) is also the LV model of the system. The LV matrices of the system are the following:

$$M = \begin{bmatrix} 0 & 1 \\ -1 & 2k_1 \end{bmatrix}, \quad N = \begin{bmatrix} -2 \\ 2(\delta_1 - 1) \end{bmatrix}$$

It is noticeable that matrix M is not upper triangular (i.e., the equilibrium cannot be manipulated partially). However, with a fortunate choice of δ_1 (e.g., $\delta_1 = 2.5$) one can modify the value of the (nonwashout-type) equilibrium of system (6.22). It is important to note that in this case the equilibrium will be positive, but one cannot decide its value.

The other free parameter (k_1) can be used for stabilizing this equilibrium. So, k_1 and the two parameters of the Lyapunov function are given to the iterative LMI (ILMI) algorithm [34]. It gives the following results:

$$k_1 = -0.0013, \quad C = \begin{bmatrix} 1.2822 & 0 \\ 0 & 1.2822 \end{bmatrix}$$

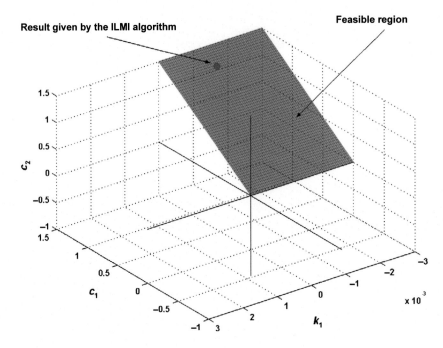

Fig. 6.17 BMI feasibility region for the example in Section 6.3.2.

Fig. 6.17 shows the feasibility region of the globally stabilizing BMI problem and the solution given by the ILMI algorithm [40]. The obtained feedback with parameters k_1 and δ_1 globally stabilizes the system in the positive orthant. Indeed, the closed-loop system has a unique equilibrium state

$$\left[\,\bar{x}_1 \bar{x}_2\, \right] = \left[\begin{array}{c} 2.9948 \\ 2.0000 \end{array} \right]$$

in the positive orthant $\mathbb{R}^2_+$, for which the locally linearized system matrix has eigenvalues with strictly negative real part; this way at least local stability can be proved for the equilibrium.

6.3.3 Fully Actuated Fermentation Example in QP Form
The second fermentation example is the same fermentation process examined in Section 3.4, but this time, biomass is also fed to the reactor with manipulable inlet concentration X_F.

The state equations with x_1 being the biomass and x_2 being the substrate concentrations are as follows:

$$\dot{x}_1 = \mu_{max} x_1 x_2 + \frac{F}{V}(X_F - x_1)$$
$$\dot{x}_2 = -\frac{\mu_{max}}{Y} x_1 x_2 + \frac{F}{V}(S_F - x_2) \tag{6.23}$$

The parameters of the system are the same as in the previous case (seen in Table 6.3).

The QP form of the model is

$$\dot{x}_1 = x_1 \left(x_2 + 2x_1^{-1} X_F - 2 \right)$$
$$\dot{x}_2 = x_2 \left(-x_1 + 2x_2^{-1} S_F - 2 \right) \tag{6.24}$$

The manipulable inputs are X_F and S_F.

For fixed values of the input concentrations $X_F^* = 0$ and $S_F^* = 1$, the system has no equilibrium in the strictly positive orthant, but has one asymptotically stable washout equilibrium on the boundary

$$\begin{bmatrix} x_1^* \\ x_2^* \\ X_F^* \\ S_F^* \end{bmatrix} = \begin{bmatrix} 0 \\ 1 \\ 0 \\ 1 \end{bmatrix}$$

In order to stabilize the system at a positive steady-state (equilibrium) point, the feedback structure is chosen to be

$$X_F = k_1 x_1^2 + \delta_1 x_1$$
$$S_F = k_2 x_2^2 + \delta_2 x_2$$

Parameters k_1 and k_2 are for stabilizing the system, δ_1 and δ_2 will be used to shift the equilibrium.

The closed-loop system is

$$\dot{x}_1 = x_1 \left(2(\delta_1 - 1) + x_2 + 2k_1 x_1 \right)$$
$$\dot{x}_2 = x_2 \left(2(\delta_2 - 1) - x_1 + 2k_2 x_2 \right)$$

that is also the LV form of the closed-loop system model.

The iterative BMI algorithm yielded the following parameters for the feedback and the Lyapunov function:

$$k_1 = -1.0004, \quad k_2 = -1.0004, \quad C = \begin{bmatrix} 1.0004 & 0 \\ 0 & 1.0004 \end{bmatrix}$$

We would like to prescribe a strictly positive equilibrium instead of the original one. Suppose that the desired equilibrium is at

$$\begin{bmatrix} \tilde{x}_1 \\ \tilde{x}_2 \end{bmatrix} = \begin{bmatrix} 0.5 \\ 0.5 \end{bmatrix}$$

Expressing the values of δ_1 and δ_2 from the state equations in which the desired equilibrium point is substituted in yields

$$\delta_1 = 1.2502, \quad \delta_2 = 1.7502$$

Indeed, the closed-loop system with the determined parameters $k_1, k_2, \delta_1, \delta_2$ has an asymptotically stable equilibrium point $[\tilde{x}_1, \tilde{x}_2]^T$.

It is apparent that in this example, with a higher degree of freedom it was possible to shift the steady-state point of the system to a desired value.

6.3.4 Feedback Design for a Simple Fermentation Process

This example presents a fermentation process similar to the one described by Eq. (3.48), just the reaction kinetics (i.e., function $\mu(x_2)$) is different. This one is a monotonous function of the substrate concentration x_2, that results in a simpler nonlinearity.

The system is described by the non-QP input-affine state-space model

$$\dot{x}_1 = \mu(x_2)x_1 + \frac{(X_F - x_1)F}{V}$$
$$\dot{x}_2 = -\frac{\mu(x_2)x_1}{Y} + \frac{(S_F - x_2)F}{V} \qquad (6.25)$$
$$\mu(x_2) = \mu_{max}\frac{x_2}{K_S + x_2}$$

where the inlet substrate and biomass concentrations denoted by S_F and X_F are the manipulated inputs. The variables and parameters of the model together with their units and parameter values are given in Table 6.4.

The system has a unique locally stable equilibrium point in the positive orthant

$$\begin{bmatrix} \bar{x}_1 \bar{x}_2 \end{bmatrix} = \begin{bmatrix} 0.6500 \\ 0.4950 \end{bmatrix} \qquad (6.26)$$

with steady-state inputs

$$\begin{bmatrix} \bar{X}_F \bar{S}_F \end{bmatrix} = \begin{bmatrix} 0.6141 \\ 4.3543 \end{bmatrix}$$

Table 6.4 Variables and Parameters of the Fermenter Model (6.25)

x_1	Biomass concentration		g/L
x_2	Substrate concentration		g/L
S_F	Substrate feed concentration		g/L
X_F	Biomass feed concentration		g/L
F	Inlet feed flow-rate	1.0000	L/h
V	Volume	97.8037	L
Y	Yield coefficient	0.0097	–
μ_{max}	Kinetic parameter	0.0010	L/h
K_S	Kinetic parameter	0.5	L/g

By introducing a new differential variable $\eta = \frac{1}{K_S + x_2}$, one arrives at a third-differential equation

$$\dot{\eta} = -\frac{1}{(K_S + x_2)^2} \cdot \frac{dx_2}{dt} = -\eta^2 \cdot \left(-\frac{\mu_{max}}{Y} x_1 x_2 \eta + \frac{(S_F - x_2)F}{V} \right)$$

$$= \eta \left(\frac{\mu_{max}}{Y} x_1 x_2 \eta^2 + \frac{F}{V} x_2 \eta - S_F \frac{F}{V} \eta \right)$$

(6.27)

that completes the ones for x_1 and x_2.

Thus the original system (6.25) can be represented by the following three QP differential equations:

$$\dot{x}_1 = x_1 \cdot \left(-\frac{F}{V} + \mu_{max} x_2 \eta + \frac{F}{V} x_1^{-1} X_F \right)$$

$$\dot{x}_2 = x_2 \cdot \left(-\frac{F}{V} - \frac{\mu_{max}}{Y} x_1 \eta + \frac{F}{V} x_2^{-1} S_F \right)$$

$$\dot{\eta} = \eta \cdot \left(\frac{F}{V} x_2 \eta + \frac{\mu_{max}}{Y} x_1 x_2 \eta^2 - \frac{F}{V} \eta S_F \right)$$

Using a wise choice of the feedback structure, the quasimonomials of the closed-loop system may decrease. In our case the feedback structure is chosen to be

$$X_F = k_1 x_1 x_2 \eta + \delta_1 x_1$$
$$S_F = k_2 x_1 x_2 \eta + \delta_2 x_2$$

The closed-loop QP system is then

$$\dot{x}_1 = x_1 \cdot \left(-\frac{F}{V} + \left(\mu_{max} + k_1 \frac{F}{V} \right) x_2 \eta \right)$$

$$\dot{x}_2 = x_2 \cdot \left(-\frac{F}{V} + \left(-\frac{\mu_{max}}{Y} + k_2 \frac{F}{V} \right) x_1 \eta \right)$$

$$\dot{\eta} = \eta \cdot \left(\frac{F}{V} x_2 \eta + \left(\frac{\mu_{max}}{Y} - k_2 \frac{F}{V} \right) x_1 x_2 \eta^2 \right)$$

Note that for the globally stabilizing feedback design phase, parameters δ_1 and δ_2 are set to zero, since they will be used for shifting the equilibrium of the closed-loop system to the original fermenter's one. It is apparent that the closed-loop system has only three quasimonomials: $x_2 \eta$, $x_1 \eta$, and $x_1 x_2 \eta^2$.

The solution of the BMI problem gives the feedback gain parameters

$$k_1 = -1.5355$$
$$k_2 = 43.6516$$

which make the system globally asymptotically stable (in the positive orthant) with the Lyapunov function (4.4) having parameters:

$$c_1 = 0.0010, \quad c_2 = 0.0761, \quad c_3 = 0.0760$$

The equilibrium (6.26) of the open-loop fermenter can be reset by expressing δ_1 and δ_2 from the steady-state equations. This gives $\delta_1 = 1.7152$, $\delta_2 = -20.9293$, so the equilibrium point (6.26) of the fermentation process (6.25) is globally stabilized.

APPENDIX A

Notations and Abbreviations

A.1 NOTATIONS

GENERALLY USED NOTATIONS

A state matrix of a linear time invariant state-space model

B input matrix of a linear time invariant state-space model

C output matrix of a linear time invariant state-space model

D feedforward matrix of a linear time invariant state-space model

u input variable of a dynamical system

y output variable of a dynamical system

x state variable of a dynamical system

$\mathbb{N}$ the set of integer numbers

$\overline{\mathbb{N}}_{+}^{n}$ the set of nonnegative integer numbers

$\mathbb{R}$ the set of real numbers

$\mathbb{R}^{n}$ n-dimensional Euclidean space

$\mathbb{R}_{+}^{n}$ n-dimensional positive orthant

$\overline{\mathbb{R}}_{+}^{n}$ n-dimensional nonnegative orthant

v_{j} jth element of a vector v

W_{ij} **or** $W_{i,j}$ (i,j)th element of a matrix W (indexing order: row, column)

$W_{i,\cdot}$ ith row of a matrix W

$W_{\cdot j}$ jth column of a matrix W

$\lambda_{i}[W]$ the ith eigenvalue of a square matrix W

$W > 0$ ($W \geq 0$) the symmetric matrix W is positive definite (positive semidefinite)

$W < 0$ ($W \leq 0$) the symmetric matrix W is negative definite (negative semidefinite)

$W \prec 0$ ($W \preceq 0$) the entries of matrix W are negative (nonpositive)

$W \succ 0$ ($W \succeq 0$) the entries of matrix W are positive (nonnegative)

$\underline{0}$ vector with all entries equal to 0

$\underline{1}$ vector with all entries equal to 1

I unit matrix with appropriate dimension

$\mathbf{0}$ matrix with zero entries (zero matrix) with appropriate dimension

$\mathbf{diag}\{z\}$ $n \times n$ diagonal matrix containing the entries of vector $z \in \mathbb{R}^n$ in its diagonal

$\mathbf{Ln}(z)$ element-wise logarithm of a positive vector z

$\mathbf{Tr}(Z)$ trace of a quadratic matrix Z

$\dot{x}$ or $\frac{d}{dt}x$ time derivative of the signal $x(t)$

$L_f h(x)$ Lie-derivative of f along h

NOTATIONS FOR QUASIPOLYNOMIAL AND LOTKA-VOLTERRA SYSTEMS

A $n \times m$ coefficient matrix of the QP system

$\tilde{A}$ $n \times (m+1)$ coefficient matrix of the time-rescaled QP system

B $m \times n$ exponent matrix of the QP system

$\tilde{B}$ $(m+1) \times n$ exponent matrix of the time-rescaled QP system

C $m \times m$ diagonal matrix containing the coefficients of V: $C = \mathrm{diag}(c_i)$, $i = 1,\ldots,m$

λ coefficient vector of the QP system $\lambda \in \mathbb{R}^n$

Λ coefficient vector of the LV system $\Lambda \in \mathbb{R}^m$

$\mathcal{M}$ $m \times m$ coefficient matrix of the LV system ($\mathcal{M} = \mathcal{B}\mathcal{A}$)

$\tilde{M}$ $(m+1) \times (m+1)$ LV coefficient matrix of the time-rescaled system

p vector of quasimonomials, $p \in \mathbb{R}^m$

V entropy-like Lyapunov function: $V(p) = \sum_{i=1}^{m} c_i \left(p_i - p_i^* - p_i^* \ln \frac{p_i}{p_i^*} \right)$

Ω $n \times 1$ row vector containing the parameters of time-rescaling

x n-dimensional state variable vector

x^* equilibrium state variable

NOTATIONS FOR KINETIC SYSTEMS

A_k $m \times m$ Kirchhoff (or kinetics) matrix

α, β positive stoichiometric coefficients

C m-dimensional vector of chemical complexes

$C_i \to C_j$ reaction involving reactant (source) complex C_i and product complex C_j

$\mathcal{C}$ the set of complexes

δ integer optimization variables

k_j reaction rate coefficient corresponding to the jth reaction

k_{ij} reaction rate coefficient corresponding to reaction $C_i \to C_j$

$\mathcal{N}$ $n \times r$ stoichiometric matrix of a reversible reaction network

$\mathcal{R}$ the set of reactions

ρ_j rate of the jth reaction

ρ_{ij} rate of the reaction corresponding to reaction $C_i \to C_j$

$\psi \in \mathbb{R}^n \mapsto \mathbb{R}^m$ monomial function of the kinetic system: $\psi_j(x) = \prod_{i=1}^{n} x_i^{Y_{ij}}$

$\mathcal{S}$ the set of species

S stoichiometric matrix

S stoichiometric subspace of a reaction network

$\Sigma = (\mathcal{S}, \mathcal{C}, \mathcal{R})$ kinetic system characterized by the triplet $(\mathcal{S}, \mathcal{C}, \mathcal{R})$

$\Sigma = (Y, A_k)$ kinetic system characterized by the pair (Y, A_k)

X n-dimensional vector of species

$[[X_i]]$ concentration of species X_i

x n-dimensional vector of species concentrations (state variable, $x = [X]$)

x^* equilibrium concentration vector

Y $n \times m$ complex composition matrix

y continuous optimization variables

V entropy-like Lyapunov function: $V(x) = \sum_{i=1}^{n} x_i \left(\ln \left(\frac{x_i}{x_i^*} - 1 \right) \right) + x_i^*$

Mathematical Tools

The aim of this chapter is to briefly outline the mathematical and computational background for the all parts of the book. We give some basic information on graph theory to support the description of kinetic systems. Then the matrices of key importance used in the book will be described. Finally, the optimization-based computational background of model analysis and controller synthesis will be outlined.

B.1 DIRECTED GRAPHS

It is often useful to assign directed graphs to dynamical models to study their properties [73]. Therefore, we give the essential notions on graphs, particularly on directed graphs in this section.

An *undirected graph*, or simply *graph*, is an ordered pair $\mathcal{G} = (V, E)$, where V is a finite set containing the *vertices* or *nodes*, and $E \subseteq \{\{v_1, v_2\} | v_1, v_2 \in V\}$ is the set of *edges*. If $e = \{v_1, v_2\} \in E$, then nodes v_1 and v_2 are connected in the graph by the undirected edge e. This means that in the case of undirected graphs, the edges can be considered "unordered pairs."

A *directed graph* $G = (V, E)$, where V is a finite set containing the *vertices* or *nodes*, and $E \subseteq \{(v_1, v_2) | v_1, v_2 \in V\}$ is the set of *directed edges*. For $e = (v_1, v_2) \in E$, v_1 is called the *tail* and v_2 is called the *head* of the directed edge e. Directed edges are usually represented as arrows, that is, $e = (v_1, v_2)$ is depicted as $v_1 \longrightarrow v_2$.

Sometimes it is practical to assign weights (in our case, real numbers) to the edges of graphs. In such a case, we speak about a *weighted (directed) graph*. The following notions will be related to directed graphs, since we will mostly use those for characterizing certain properties of the studied models.

A directed graph $G' = (V', E')$ is called a *subgraph of* $G = (V, E)$ if $V' \subseteq V$ and $E' \subseteq E$.

A *walk* in a directed graph is a sequence $W = v_1 v_2 \ldots v_{k-1} v_k$, where $v_i \in V$ for $i = 1, \ldots, k$, and $(v_i, v_{i+1}) \in E$ for $i = 1, \ldots, k-1$. W is called a *directed path* if all of its vertices are distinct. A directed path W is called a *directed cycle* if the vertices $v_1, v_2, \ldots, v_{k-1}$ are distinct, $k \geq 3$ and $v_1 = v_k$. A directed graph is *acyclic* if it does not contain a directed cycle. If each strongly connected component of a directed graph G is contracted to a single vertex, the resulting directed graph is a directed acyclic graph. The sequence $S = v_1 v_2 \ldots v_{k-1} v_k$ is a *semipath* if all of its vertices are distinct, and either $(v_i, v_{i+1}) \in E$ and/or $(v_{i+1}, v_i) \in E$ for $i = 1, \ldots, k-1$. If there is a directed path from v_1 to v_2, then v_2 is said to be *reachable* from v_1. The directed graph G is called *strongly connected* if each vertex of G is reachable from each other vertex. It is worth mentioning that a directed graph is strongly connected if and only if each of its directed edges lies on a directed cycle. G is said to be *weakly connected* if there exists a semipath from each vertex of G to each other vertex. A *strong (weak) component* of G is a maximal strongly (weakly) connected subgraph of G. We remark that by definition, a single vertex is trivially connected to itself. Therefore, a strongly connected component may consist of one single vertex.

The so-called traps in directed graphs have an important role in analyzing biological systems and the properties of certain related matrices. A *simple trap* is a strongly connected component from which there is no outgoing directed edge.

Example 18 (Reaction graph of the yeast G-protein cycle). The basic graph-related notions applied to directed graphs are shown in this example on the reaction graph of the yeast G-protein cycle using the model published in [69]. The model involves a so-called heterotrimeric G-protein containing three different subunits. In response to the extracellular ligand binding, the protein dissociates to G-α and G-$\beta\gamma$ subunits, where the active and inactive forms of the G-α subunit can also be distinguished. The reaction network model involves the following species: R and L represent the receptor and the corresponding ligand, respectively; RL refers to the ligand-bound receptor; G is the G-protein located on the intracellular membrane surface; G_a and G_d denote the active and the inactive forms of the G-α subunit; and G_{bg} is the G-$\beta\gamma$ subunit.

The reaction graph of the model is shown in Fig. B.1. The strong components are framed by dashed rectangles that are numbered from 1 to 8. Strong components 2, 4, 6, and 7 are traps. The reaction graph consists of

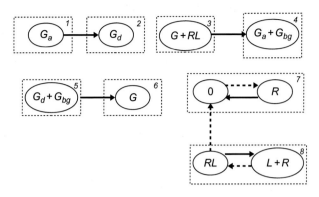

Fig. B.1 Reaction graph corresponding to Example 18 for the illustration of basic graph theoretical notions.

four weak components (linkage classes). Strong components 7 and 8 contain directed cycles. A directed path of length 3 is indicated by dashed arrows from the complex $R + L$ to R.

B.2 MATRICES OF KEY IMPORTANCE: METZLER MATRICES, COMPARTMENTAL MATRICES, KIRCHHOFF MATRICES, AND THEIR MOST IMPORTANT PROPERTIES

In this section, some matrix types and their properties will be introduced that will play a key role in the analysis of the introduced dynamical system classes. We will assume in all definitions that M is an $n \times n$ real matrix.

A matrix M is called *symmetric* or *skew-symmetric* if $M = M^T$ or $M = -M^T$, respectively. It is known that real symmetric matrices have real eigenvalues, and the nonzero eigenvalues of skew-symmetric matrices are purely imaginary. It is also easy to see that any real quadratic matrix M can be written as a sum of a symmetric (M_1) and a skew-symmetric (M_2) matrix as $M = M_1 + M_2$, where

$$M_1 = \frac{M + M^T}{2} \quad \text{and} \quad M_2 = \frac{M - M^T}{2} \tag{B.1}$$

A symmetric matrix M is called *positive definite* (*positive semidefinite*) if

$$x^T M x > 0 \, (x^T M x \geq 0), \quad \forall x \in \mathbb{R}^n \tag{B.2}$$

Similarly, a symmetric matrix M is called *negative definite* (*negative semidefinite*) if

$$x^T M x < 0 (x^T M x \leq 0), \quad \forall x \in \mathbb{R}^n \tag{B.3}$$

A symmetric matrix M is called *indefinite* if there exist $x_1, x_2 \in \mathbb{R}^n$ such that

$$x_1^T M x_1 < 0 < x_2^T M x_2 \tag{B.4}$$

It is easy to computationally check the definiteness of a matrix. A symmetric matrix M is positive definite (positive semidefinite) if and only if it has positive (nonnegative) eigenvalues. A symmetric matrix M is negative definite (negative semidefinite) if and only if it has negative (nonpositive) eigenvalues. Clearly, indefinite matrices have both positive and negative eigenvalues. For a symmetric matrix M, $M < 0$, $M \leq 0$, $M > 0$, and $M \geq 0$ denote negative definiteness, negative semidefiniteness, positive definiteness, and positive semidefiniteness, respectively.

A square matrix is called *irreducible* if it cannot be transformed into block upper-triangular form using simultaneous row and column permutations. A matrix M is irreducible if and only if the directed graph G_M, the directed graph of matrix M (see Eq. (B.10) later for the definition) is strongly connected.

A matrix M is called a *Hurwitz stable matrix* or simply *Hurwitz matrix* if the real parts of its eigenvalues are strictly negative, that is, $\lambda_i[M] < 0$ for $i = 1, \ldots, n$. We recall that the eigenvalues of M are the roots of its characteristic polynomial. It is also useful to characterize the Hurwitz property of a matrix as follows. A matrix M is Hurwitz if and only if for any $n \times n$ positive definite symmetric matrix Q, there exists an $n \times n$ positive definite symmetric matrix P such that

$$M^T P + PM = -Q \tag{B.5}$$

Eq. (B.5) is called the *Lyapunov equation* for M. We call M a *diagonally stable* matrix, if there exists a diagonal solution P for the Lyapunov equation (B.5), that is, there exists a positive definite diagonal P such that $M^T P + PM < 0$ [74]. We say that a matrix M is D-stable if the product DA is Hurwitz stable for any $n \times n$ positive definite diagonal matrix D.

We say that a square matrix M is a *Metzler matrix* if all of its nondiagonal entries are nonnegative, that is, $M_{ij} \geq 0$ for all $i \neq j$. Metzler matrices have an important role in characterizing nonnegative and compartmental systems. First of all, the linear system (2.3) is nonnegative if and only if A is a Metzler matrix [75].

A special subset of Metzler matrices is the class of compartmental matrices: M is said to be a *compartmental matrix* if its off-diagonal entries are nonnegative and its column-sums are nonpositive, that is,

$$M_{ij} \geq 0, \quad i,j = 1,\ldots,n, \; i \neq j \tag{B.6}$$

$$\sum_{i=1}^{n} M_{ij} \leq 0, \quad j = 1,\ldots,n \tag{B.7}$$

It follows from Eqs. (B.6), (B.7) that the diagonal elements of a compartmental matrix are nonpositive.

Specializing Metzler and compartmental matrices further, we say that M is a *Kirchhoff matrix* if its off-diagonal entries are nonnegative and its column sums are exactly zero, that is

$$M_{ij} \geq 0, \quad i,j = 1,\ldots,n, \; i \neq j \tag{B.8}$$

$$\sum_{i=1}^{n} M_{ij} = 0, \quad j = 1,\ldots,n \tag{B.9}$$

Because of the zero column sums, M is also a so-called *column conservation matrix*. It is visible from the linear dependence of the rows shown by Eq. (B.9) that a Kirchhoff matrix cannot be of full rank, that is, it has at least one zero eigenvalue.

We can assign a directed graph to any quadratic matrix as follows. For any $M \in \mathbb{R}^{n \times n}$, the directed graph $G_M = (V, E)$ is defined as:

$$V = \{1,\ldots,n\}, \quad E = \{(i,j) \mid M_{ji} \neq 0, \; i,j \in V, \; i \neq j\} \tag{B.10}$$

that is, the nodes of the graph are the indices of the rows/columns, and a directed edge is assigned to each nonzero off-diagonal entry of the matrix.

The following properties of the earlier-mentioned matrices are useful for the dynamical analysis of the related models:

(P1) The Metzler matrix M is Hurwitz stable if and only if it is nonsingular and $M^{-1} \preceq 0$.

(P2) The Metzler matrix M is Hurwitz stable if and only if there exists a vector $v \in \mathbb{R}_+^n$ such that $v^T M \prec 0$.

(P3) If a Metzler matrix M is Hurwitz stable, then it has strictly negative diagonal entries.

(P4) Hurwitz stable Metzler matrices are diagonally stable [76].

(P5) Diagonally stable matrices are D-stable [74].

(P6) Let $M_1, M_2 \in \mathbb{R}^{n \times n}$ Hurwitz stable Metzler matrices. Then $M_1 + \delta M_2$ is Hurwitz stable for all $\delta > 0$ if and only if $H_1 + \delta H_2$ is nonsingular for all $\delta > 0$ [77].

(P7) The eigenvalues of compartmental matrices are either zero or they have negative real parts. This trivially implies that compartmental matrices cannot have purely imaginary eigenvalues or eigenvalues with positive real parts.

(P8) Compartmental matrices of full rank are Hurwitz stable and therefore, diagonally stable.

We remark that (P2) is a simple convex constraint that can be computationally checked using, for example, linear programming (LP).

There are strong results on the Hurwitz stability of Metzler matrices even if the matrix elements are not precisely known. Two matrices $M_1, M_2 \in \mathbb{R}^{n \times m}$ are called *structurally equal* if the positions of the zero and nonzero elements are the same in M_1 and M_2, that is, $[M_1]_{i,j} \neq 0$ if and only if $[M_2]_{i,j} \neq 0$. A Metzler matrix M is called *sign-stable*, if all Metzler matrices that are structurally equal to M are Hurwitz stable. A Metzler matrix M is sign stable if and only if the directed graph G_M is acyclic. Another equivalent condition for the sign stability of a Metzler matrix M is that there exists a vector $v \in \mathbb{R}_+^n$ such that $v^T \mathrm{sgn}(M) \prec 0$ where $\mathrm{sgn}(M)$ is the sign version of the matrix M.

B.3 BASICS OF THE APPLIED COMPUTATIONAL TOOLS

In this section, we briefly summarize those optimization tools that are used in the book for CRN structure computation and controller design. The summary is taken from [78].

B.3.1 Linear Programming

LP is an optimization technique, where a linear objective function is minimized/maximized subject to linear equality and inequality constraints [79, 80]. Beside numerous other fields, LP has been widely used in chemistry and chemical engineering in the areas of system analysis [81, 82], simulation [83], and design [84]. The standard form of LP problems that is used in this book is the following

$$\text{minimize} \quad c^T y \tag{B.11}$$

$$\text{subject to:} \quad Ay = b \tag{B.12}$$

$$y \geq 0 \tag{B.13}$$

where $y \in \mathbb{R}^n$ is the vector of decision variables, $c \in \mathbb{R}^n$ $A \in \mathbb{R}^{p \times n}$, $b \in \mathbb{R}^p$ are known vectors and matrices, and "$\geq$" in Eq. (B.13) means element-wise nonstrict inequality.

The feasibility of the simple LP problem (B.11)–(B.13) can be checked, for example, using the following necessary and sufficient condition.

Theorem 3 (Dantzig and Thapa [79]). *Consider the auxiliary LP problem*

$$\text{minimize} \sum_{i=1}^{p} z_i \tag{B.14}$$

$$\text{subject to: } Ax + z = b \tag{B.15}$$

$$x \geq 0, z \geq 0 \tag{B.16}$$

where $z \in \mathbb{R}^p$ is a vector of auxiliary variables. There exists a feasible solution for the LP problem (B.11)–(B.13) if and only if the auxiliary LP problem (B.14)–(B.16) has optimal value 0 with $z_i = 0$ for $i = 1, \ldots, p$.

It is easy to see that $x = 0$, $z = b$ is always a feasible solution for Eqs. (B.14)–(B.16). The previous theorem will be useful for establishing that no feasible solution for a pure LP problem exists.

B.3.2 Mixed Integer LP and Propositional Logic

A mixed integer linear program is the maximization or minimization of a linear function subject to linear constraints that may have both real-valued and integer-valued variables. A mixed integer linear program with k variables (denoted by $y \in \mathbb{R}^k$) and p constraints can be written as [85]:

$$\text{minimize } c^T y$$
$$\text{subject to: } A_1 y = b_1$$
$$A_2 y \leq b_2 \tag{B.17}$$
$$l_i \leq y_i \leq u_i \text{ for } i = 1, \ldots, k$$
$$y_j \text{ is integer for } j \in I, \ I \subset \{1, \ldots, k\}, \ I \neq \emptyset$$

where $c \in \mathbb{R}^k$, $A_1 \in \mathbb{R}^{p_1 \times k}$, $A_2 \in \mathbb{R}^{p_2 \times k}$, and $p_1 + p_2 = p$.

If all the variables can be real, then Eq. (B.17) is a simple LP problem that can be solved in polynomial time. However, if any of the variables is integer, then the problem generally becomes NP-hard. In spite of this, there exist a number of free (e.g., YALMIP [86] or the GNU Linear Programming Kit [87]) and commercial (such as CPLEX or TOMLAB [88]) solvers that can efficiently handle many practical problems.

It is well known that statements in propositional calculus can be transformed into linear inequalities, as shown briefly following. The notations of the following summary are mostly from [89]. A statement, such as $x \leq 0$ that can have a truth value of "T" (true) or "F" false is called a *literal* and will be denoted by S_i. In Boolean algebra, literals can be combined into *compound statements* using the following *connectives*: "$\wedge$" (and), "$\vee$" (or), "$\sim$" (not), "$\rightarrow$" (implies), "$\leftrightarrow$" (if and only if), and "$\oplus$" (exclusive or). The truth table for the previously listed connectives is given in Table B.1.

A propositional logic problem, where a statement S_1 must be proved to be true given a set of compound statements containing literals $S_1, \ldots, S_n$, can be solved by means of a linear integer program. For this, logical variables denoted by δ_i ($\delta_i \in \{0, 1\}$) must be associated with the literals S_i. Then, the original compound statements can be translated to linear inequalities involving the logical variables δ_i. A list of equivalent compound statements and linear equalities or inequalities taken from [90] is shown in Table B.2.

Table B.1 Truth Table

S_1	S_2	$\sim S_1$	$S_1 \vee S_2$	$S_1 \wedge S_2$	$S_1 \rightarrow S_2$	$S_1 \leftrightarrow S_2$	$S_1 \oplus S_2$
T	T	F	T	T	T	T	F
T	F	F	T	F	F	F	T
F	T	T	T	F	T	F	T
F	F	T	F	F	T	T	F

Table B.2 Equivalent Compound Statements and Linear Equalities/Inequalities

Compound Statement	Equivalent Linear Equality/Inequality
$S_1 \vee S_2$	$\delta_1 + \delta_2 \geq 1$
$S_1 \wedge S_2$	$\delta_1 = 1, \delta_2 = 1$
$\sim S_1$	$\delta_1 = 0$
$S_1 \rightarrow S_2$	$\delta_1 - \delta_2 \leq 0$
$S_1 \leftrightarrow S_2$	$\delta_1 - \delta_2 = 0$
$S_1 \oplus S_2$	$\delta_1 + \delta_2 = 1$

B.3.3 Linear and Bilinear Matrix Inequalities

A (nonstrict) linear matrix inequality (LMI) is an inequality of the form

$$F(x) = F_0 + \sum_{i=1}^{m} x_i F_i \geq 0 \qquad (B.18)$$

where $x \in \mathbb{R}^m$ is the variable and $F_i \in \mathbb{R}^{n \times n}$, $i = 0, \ldots, m$ are given symmetric matrices. The inequality symbol in Eq. (B.18) stands for the positive semidefiniteness of $F(x)$.

One of the most important properties of LMIs is the fact that they form a convex constraint on the variables, that is, the set $\{x \mid F(x) \geq 0\}$ is convex. LMIs have been playing an increasingly important role in the field of optimization and control theory since a wide variety of different problems (linear and convex quadratic inequalities, matrix norm inequalities, convex constraints, etc.) can be written as LMIs, and there are computationally stable and effective (polynomial time) algorithms for their solution [39, 91].

A bilinear matrix inequality (BMI) is a diagonal block composed of q matrix inequalities of the following form

$$G_0^i + \sum_{k=1}^{p} x_k G_k^i + \sum_{k=1}^{p} \sum_{j=1}^{p} x_k x_j K_{kj}^i \leq 0, \quad i = 1, \ldots, q \qquad (B.19)$$

where $x \in \mathbb{R}^p$ is the decision variable to be determined and G_k^i, $k = 0, \ldots, p$, $i = 1, \ldots, q$ and K_{kj}^i, $k, j = 1, \ldots, p$, $i = 1, \ldots, q$ are symmetric, quadratic matrices.

The main properties of BMIs are that they are nonconvex in x (which makes their solution numerically much more complicated than that of linear matrix inequalities), and their solution is NP-hard [37]. However, there exist practically applicable and effective algorithms for BMI solution [40, 92].

B.4 BASIC NOTIONS FROM SYSTEMS AND CONTROL THEORY

In this book, we only use the class of concentrated parameter nonlinear systems, that have smooth nonlinear mappings in their dynamic model equations. Therefore, the most important concepts and results about their state-space models, realizations, stability, and stabilizing feedback design are briefly summarized here for the readers' convenience based on the book

[70], where more detailed information can be found. A process systems focused treatment is given in [48].

B.4.1 Input-Affine Nonlinear State-Space Models and Their Realizations

A special subclass of smooth nonlinear systems is formed by the so-called input-affine systems that have a state-space model in the following general form

$$\frac{dx}{dt}(t) = f(x(t)) + \sum_{i=1}^{m} g_i(x(t))u_i(t) \quad \text{(state equation)}$$

$$y(t) = h(x(t)) \quad \text{(output equation)} \tag{B.20}$$

with the state, input, and output vector-valued signals x, u, and y, and with the smooth nonlinear mappings

$$f: \mathbb{R}^n \mapsto \mathbb{R}^p, \quad g_i: \mathbb{R}^n \mapsto \mathbb{R}, \quad h: \mathbb{R}^n \mapsto \mathbb{R}^p$$

Furthermore, we assume that

$$f(0) = 0, \quad h(0) = 0 \tag{B.21}$$

hold.

It is important to observe that the *input signals enter into the input-affine nonlinear state-space model in a linear way*, that is, the mapping on the right-hand side of the nonlinear state-space model (B.20) is linear with respect to u.

B.4.1.1 Transformation of States

A nonlinear change of coordinates is written as

$$z = \Phi(x) \tag{B.22}$$

where Φ represents an $\mathbb{R}^n$-valued function of n variables, that is,

$$\Phi(x) = \begin{bmatrix} \phi_1(x) \\ \phi_2(x) \\ \vdots \\ \phi_n(x) \end{bmatrix} = \begin{bmatrix} \phi_1(x_1,\ldots,x_n) \\ \phi_2(x_1,\ldots,x_n) \\ \vdots \\ \phi_n(x_1,\ldots,x_n) \end{bmatrix} \tag{B.23}$$

with the following properties:

1. Φ is invertible, that is, there exists a function Φ^{-1} such that $\Phi^{-1}(\Phi(x)) = x$ for all x in $\mathbb{R}^n$.

2. Φ and Φ^{-1} are both smooth mappings, that is, they have continuous partial derivatives of any order.

A transformation of this type is called a *global diffeomorphism* on $\mathbb{R}^n$.

It is important to note that only the state variable x is transformed using the transformation (B.22); therefore, the resulting state-space model will be input-affine, and will have the same input-output dynamical properties, as the original one. Therefore, we call the original and the transformed state-space model *two equivalent realizations* of the same nonlinear system.

B.4.2 Stability Analysis

Although there exists a number of useful stability notions for nonlinear systems, we only use the notion of asymptotic stability in this book. For this purpose the notion of *truncated state equation*, that is, the special case of the state equation in Eq. (B.20) is considered

$$\frac{dx}{dt} = f(x), \quad x(0) = x_0 \tag{B.24}$$

In the case of input-affine nonlinear systems, we assume that $f(0) = 0$. Then Eq. (B.24) has a *stationary solution* $x^0(t) \equiv 0$ for $x^0(t_0) = 0$, that is also called a *steady-state point* $x^* = 0$.

The steady-state point x^* of Eq. (B.24) is *locally asymptotically stable* if $\|x(t)\| \to 0$ when $t \to \infty$ provided that

$$\|x(t_0)\| \le \epsilon \tag{B.25}$$

where ϵ is small enough.

The *domain of attraction* (abbreviated as the *DOA*) of the locally asymptotically stable steady-state point x^* is a region around it determined by the parameter ϵ in Eq. (B.25), where the condition $\|x(t)\| \to 0$ when $t \to \infty$ holds. If the condition holds on the entire domain of Eq. (B.24), then its steady-state point x^* is *globally asymptotically stable*.

B.4.2.1 Lyapunov Function, Lyapunov Theorem

The Lyapunov theorem and the Lyapunov function, the existence of which is the condition of the theorem, play a central role in analyzing asymptotic stability of nonlinear systems. This is the most widespread and almost the only generally applicable technique for this case.

A *Lyapunov function* $V[x]$ to an autonomous system with a truncated state equation $\dot{x} = f(x)$ is a scalar-valued function with the following properties:

1. Scalar function

$$V: \mathbb{R}^n \rightarrow \mathbb{R}$$

2. Positive definiteness

$$V[x(t)] > 0$$

3. Dissipativity

$$\frac{d}{dt}V[x(t)] = \frac{\partial V}{\partial x}\frac{d[x(t)]}{dt} < 0$$

The celebrated *Lyapunov theorem states that a system is asymptotically stable* (in the strong sense) *if there exists a Lyapunov function with the previous properties.*

Note that the Lyapunov criterion earlier is not constructive: it is the task of the user to find an appropriate Lyapunov function to show stability. Moreover, the reverse of the statement is *not* true.

B.4.3 Stabilizing Feedback Controllers

When we manipulate the input signals of a system in order to achieve some goal concerning its behavior then we "control" it. The most commonly applied way of controlling dynamic systems is to manipulate its behavior using *feedback*, that is, *to choose the signal value of u(t) depending on the state or output of the system.* Although there are various useful aims to manipulate the input of a system, we restrict our attention here to the only aim to stabilize it, that is called *stabilizing control.*

The simplest way is to apply a *so-called static (full) state feedback to the input-affine nonlinear state-space model* (B.20) *without or with new input* in the form

$$u = \alpha(x) \tag{B.26}$$

where $\alpha: \mathcal{X} \mapsto \mathbb{R}^m$ is a smooth function, and with the new input v in the form

$$u = \alpha(x) + \beta(x)v \tag{B.27}$$

where $\alpha: \mathcal{X} \mapsto \mathbb{R}^m$ and $\beta: \mathbb{X} \mapsto \mathbb{R}^{m \times m}$ are smooth mappings. Furthermore, $\beta(x)$ is invertible for all x and $v \in \mathbb{R}^m$ is a new vector of control variables.

B.4.4 Input-Output Linearization via State Feedback

The aim of this method is to apply a nonlinear state feedback in order to get a closed-loop system that is a linear one. Given a system in the form Eq. (B.20) where the input and output signals are both scalar, that is, the general input-affine system model (B.20) specializes to

$$\begin{aligned} \dot{x} &= f(x) + g(x)u \\ y &= h(x) \end{aligned} \tag{B.28}$$

where $f\colon \mathbb{R}^n \to \mathbb{R}^n$, $g\colon \mathbb{R}^n \to \mathbb{R}^n$, and $h\colon \mathbb{R}^n \to \mathbb{R}$. Suppose that the *relative degree* of system (B.28) is r at point x_0, that is,

1. $L_g L_f^k h(x) = 0$ for all x in a neighborhood of x_0 and $k < r - 1$
2. $L_g L_f^r h(x) \neq 0$,

where $L_f h(x)$ stands for the so-called *Lie derivative* [70]

$$L_f h(x) = \sum_{i=1}^{n} \frac{\partial h(x)}{\partial x_i} f_i(x)$$

The relative degree r equals to the number of times one has to differentiate $y(t)$ in order that $u(t)$ explicitly appear in $y^{(r)}(t)$.

Using the local coordinates transformation (B.29) in the neighborhood of x_0,

$$z = T(x) = \begin{bmatrix} h(x) \\ L_f h(x) \\ \vdots \\ L_f^{r-1} h(x) \\ \phi_1(x) \\ \vdots \\ \phi_{n-r}(x) \end{bmatrix} \tag{B.29}$$

where $\phi_i(x)$ are chosen such a way that $L_g \phi_i(x) = 0$ around x_0, $i = 1, \ldots, n - r$, and applying the input

$$u = \frac{1}{L_g L_f^{r-1} h(x)} \left(-L_f^r h(x) + v \right)$$

the system (B.28) can be rewritten in the form

$$
\begin{aligned}
\dot{z}_1 &= z_2 \\
&\vdots \\
\dot{z}_{r-1} &= z_r \\
\dot{z}_r &= v \\
\dot{z}_{r+1} &= q_{r+1}(z) \\
&\vdots \\
\dot{z}_n &= q_n(z) \\
y &= z_1
\end{aligned}
\tag{B.30}
$$

It is apparent that the first r states of Eq. (B.30) form a linear subsystem with a new input v, and there is an additional type of nonlinear dynamics, the so-called *zero dynamics* [70]. Zero dynamics describes the systems behavior when its output y is forced to be constantly zero.

The applicability condition of the earlier method is the asymptotic stability of the zero dynamics. If it holds, then any suitable control method from the area of LTI systems (e.g., an LQ type) can be applied to stabilize the linear subsystem, and the overall dynamics will be stable.

B.4.5 Uncertainty Description in Dynamic Models
Parametric uncertainties are often present in the parameters of a dynamic model because they are usually determined from experiments that are subject of measurement and estimation errors. There are two alternative but substantially different ways of describing uncertainties related to a parameter or variable in a model: the random or the set type descriptions. The set type description of uncertainties is simple: we give a set as a value of the parameter or variable within which its true value lies regardless of the different probabilities that could be associated with the individual elements of the set.

B.4.5.1 Polytopic Sets
A *polyhedron* $\mathcal{H}$ is defined as a nonempty intersection of halfspaces in $\mathbb{R}^n$, and therefore it can be given in the form

$$
\mathcal{H} = \left\{ x \in \mathbb{R}^n \mid Ax \leq b \right\}
\tag{B.31}
$$

where $A \in \mathbb{R}^{m \times n}$ and $b \in \mathbb{R}^m$.

We define a *convex polytope* $\mathcal{P}$ (which we simply call a *polytope*) as the convex hull of a nonempty finite set S of points in $\mathbb{R}^n$, that is,

$$\mathcal{P} = \left\{ \sum_{i=1}^{|S|} \lambda_i x_i \,\middle|\, \lambda_i \geq 0 \forall i = 1, \ldots, |S|, \quad \text{and} \quad \sum_{i=1}^{|S|} \lambda_i = 1 \right\} \qquad \text{(B.32)}$$

where $x_i \in \mathbb{R}^n$ for $i = 1, \ldots, |S|$. It follows from the previous definition that polytopes are bounded sets. Moreover, bounded polyhedrons and convex polytopes are equivalent, and therefore they can be represented either in the inequality form Eq. (B.31) or by using their vertex points as Eq. (B.32).

We also use *matrix polytopic sets* in this book that are defined as follows. Consider L real matrices of the same size $M^{(i)} \in \mathbb{R}^{n \times m}$ for $i = 1, \ldots, L$ and form their convex hull $\mathcal{M}$ as

$$\mathcal{M} = \left\{ \sum_{i=1}^{L} \alpha_i M^{(i)} \,\middle|\, \alpha_i \geq 0 \quad \forall i = 1, \ldots, L, \text{ and } \sum_{i=1}^{L} \alpha_i = 1 \right\} \qquad \text{(B.33)}$$

where the matrices $M^{(i)}$ for $i = 1, \ldots, L$ are called the vertex points of the set.

Now we can define a dynamic kinetic model with parametric uncertainty as a set of dynamic models in the form

$$\frac{dx(t)}{dt} = M\psi(x(t)) \qquad \text{(B.34)}$$

where the parameter matrix M is an arbitrary element of the matrix polytopic set $\mathcal{M}$, that is, $M \in \mathcal{M}$.

BIBLIOGRAPHY

[1] U. Alon, An Introduction to Systems Biology: Design Principles of Biological Circuits, Chapman & Hall, CRC, Boca Raton, FL, 2007.

[2] National Academies Press (US), Convergence: Facilitating Transdisciplinary Integration of Life Sciences, Physical Sciences, Engineering, and Beyond, 2014 (Committee on Key Challenge Areas for Convergence and Health and Board on Life Sciences and Division on Earth and Life Studies and National Research Council).

[3] S. Luther, et al., Low-energy control of electrical turbulence in the heart, Nature 475 (2011) 235–241.

[4] A.J.L. Sanz, F.J. Doyle, E. Dassau, An enhanced model predictive control for the artificial pancreas using a confidence index based on residual analysis of past predictions, J. Diabetes Sci. Technol. 11 (3) (2017) 537–544, https://doi.org/10.1177/1932296816680632.

[5] J. Sápi, L. Kovács, D.A. Drexler, P. Kocsis, D. Gajári, Z. Sápi, Tumor volume estimation and quasi-continuous administration for most effective bevacizumab therapy, PLoS ONE 10 (2015) E0142190.

[6] B. Hernández-Bermejo, V. Fairén, Lotka-Volterra representation of general nonlinear systems, Math. Biosci. 140 (1997) 1–32.

[7] S. Kingsland, Modeling Nature: Episodes in the History of Population Ecology, University of Chicago Press, Chicago, IL, 1995.

[8] B. Hernandez-Bermejo, V. Fairén, L. Brenig, Algebraic recasting of nonlinear systems of ODEs into universal formats, J. Phys. A 31 (1998) 2415–2430.

[9] L.S. Pontryagin, Ordinary Differential Equations, Addison-Wesley, Reading, MA, 1962.

[10] T. Kailath, Linear Systems, Prentice-Hall, Englewood Cliffs, NJ, 1979.

[11] W.M. Haddad, V.S. Chellaboina, Q. Hui, Nonnegative and Compartmental Dynamical Systems, Princeton University Press, Princeton, NJ, 2010.

[12] B. Hernández-Bermejo, V. Fairén, Nonpolynomial vector fields under the Lotka-Volterra normal form, Phys. Lett. A 206 (1995) 31–37.

[13] B. Hernández-Bermejo, Stability conditions and Lyapunov functions for quasi-polynomial systems, Appl. Math. Lett. 15 (2002) 25–28.

[14] A. Figueiredo, I.M. Gleria, T.M. Rocha, Boundedness of solutions and Lyapunov functions in quasi-polynomial systems, Phys. Lett. A 268 (2000) 335–341.

[15] Y. Takeuchi, Global Dynamical Properties of Lotka-Volterra Systems, World Scientific, Singapore, 1996.

[16] M. Feinberg, Chemical reaction network structure and the stability of complex isothermal reactors—I. The deficiency zero and deficiency one theorems, Chem. Eng. Sci. 42 (1987) 2229–2268.

[17] P. Érdi, J. Tóth, Mathematical Models of Chemical Reactions. Theory and Applications of Deterministic and Stochastic Models, Manchester University Press, Princeton University Press, Manchester, Princeton, 1989.

[18] N. Samardzija, L.D. Greller, E. Wassermann, Nonlinear chemical kinetic schemes derived from mechanical and electrical dynamical systems, J. Chem. Phys. 90 (1989) 2296–2304.

[19] D. Angeli, A tutorial on chemical network dynamics, Eur. J. Control 15 (2009) 398–406.

[20] M. Feinberg, F.J.M. Horn, Chemical mechanism structure and the coincidence of the stoichiometric and kinetic subspaces, Arch. Ration. Mech. Anal. 66 (1) (1977) 83–97, ISSN 0003-9527, https://doi.org/10.1007/BF00250853.

[21] V. Hárs, J. Tóth, On the inverse problem of reaction kinetics, in: M. Farkas, L. Hatvani (Eds.), Qualitative Theory of Differential Equations, Coll. Math. Soc. J. Bolyai, vol. 30, North-Holland, Amsterdam, 1981, pp. 363–379.

[22] F. Horn, Necessary and sufficient conditions for complex balancing in chemical kinetics, Arch. Ration. Mech. Anal. 49 (1972) 172–186.

[23] G. Craciun, C. Pantea, Identifiability of chemical reaction networks, J. Math. Chem. 44 (2008) 244–259.

[24] G. Szederkényi, K.M. Hangos, D. Csercsik, Computing realizations of reaction kinetic networks with given properties, in: Coping With Complexity: Model Reduction and Data Analysis, Lecture Notes in Computational Science and Engineering, Springer, 2009, pp. 253–267.

[25] A. Gábor, K.M. Hangos, J.R. Banga, G. Szederkényi, Reaction network realizations of rational biochemical systems and their structural properties, J. Math. Chem. 53 (2015) 1657–1686, https://doi.org/10.1007/s10910-015-0511-9.

[26] A. Gábor, K.M. Hangos, G. Szederkényi, Linear conjugacy in biochemical reaction networks with rational reaction rates, J. Math. Chem. 54 (2016) 1658–1676, https://doi.org/10.1007/s10910-016-0642-7.

[27] G. Lipták, G. Szederkényi, M. Hangos, Kinetic feedback design for polynomial systems, J. Process Control 41 (2016) 56–66, https://doi.org/10.1016/j.jprocont.2016.03.002.

[28] T. Plesa, T. Vejchodsky, R. Erban, Chemical reaction systems with a homoclinic bifurcation: an inverse problem, J. Math. Chem. (2016), https://doi.org/10.10007/s10910-016-0656-1.

[29] M.D. Johnston, D. Siegel, Linear conjugacy of chemical reaction networks, J. Math. Chem. 49 (2011) 1263–1282.

[30] G. Farkas, Kinetic lumping schemes, Chem. Eng. Sci. 54 (1999) 3909–3915.

[31] M.D. Johnston, D. Siegel, G. Szederkényi, A linear programming approach to weak reversibility and linear conjugacy of chemical reaction networks, J. Math. Chem. 50 (2012) 274–288, https://doi.org/10.1007/s10910-011-9911-7.

[32] R. Díaz-Sierra, B. Hernández-Bermejo, V. Fairén, Graph-theoretic description of the interplay between non-linearity and connectivity in biological systems, Math. Biosci. 156 (1999) 229–253.

[33] L. Brenig, A. Goriely, Universal canonical forms for the time-continuous dynamical systems, Phys. Rev. A 40 (1989) 4119–4122.

[34] A. Magyar, G. Szederkényi, K.M. Hangos, Globally stabilizing feedback control of process systems in generalized Lotka-Volterra form, J. Process Control 18 (2008) 80–91, https://doi.org/10.1016/j.jprocont.2007.05.003.

[35] E. Kaszkurewicz, A. Bhaya, Matrix Diagonal Stability in Systems and Computation, Springer Science & Business Media, New York, NY, 2012.

[36] G. Szederkényi, K.M. Hangos, A. Magyar, On the time-reparametrization of quasi-polynomial systems, Phys. Lett. A 334 (2005) 288–294.

[37] J.G. Van Antwerp, R.D. Braatz, A tutorial on linear and bilinear matrix inequalities, J. Process Control 10 (2000) 363–385.

[38] S.P. Boyd, L. El Ghaoui, E. Feron, V. Balakrishnan, Linear Matrix Inequalities in System and Control Theory, vol. 15, SIAM, 1994.

[39] C. Scherer, S. Weiland, Linear Matrix Inequalities in control, Lecture Notes, Dutch Institute for Systems and Control, Delft, The Netherlands 3 (2000).

[40] M. Kocvara, M. Stingl, A code for convex nonlinear and semidefinite programming, Optimization Methods and Software 8 (2003) 317–333.

[41] D.F. Anderson, A proof of the Global Attractor Conjecture in the single linkage class case, SIAM J. Appl. Math. 71 (2011) 1487–1508.

[42] M. Chaves, E.D. Sontag, State-estimators for chemical reaction networks of Feinberg-Horn-Jackson zero deficiency type, Eur. J. Control 8 (2002) 343–359.

[43] M. Chaves, Input-to-state stability of rate-controlled biochemical networks, SIAM J. Control Optim. 44 (2005) 704–727.

[44] K.M. Hangos, G. Szederkényi, The effect of conservation on the dynamics of chemical reaction networks, in: B.E. Ydstie, B. Maschke, D. Dochain (Eds.), IFAC Workshop on Thermodynamic Foundations of Mathematical Systems Theory, July 13–16, Lyon, France, 2013, pp. 30–35, https://doi.org/10.3182/20130714-3-FR-4040.00008.

[45] I. Nagy, J. Tóth, Quadratic first integrals of kinetic differential equations, J. Math. Chem. 52 (1) (2014) 93–114, https://doi.org/10.1007/s10910-013-0247-3.

[46] B. Pongrácz, G. Szederkényi, K.M. Hangos, An algorithm for determining a class of invariants in quasi-polynomial systems, Comput. Phys. Commun. 175 (2006) 204–211, https://doi.org/10.1016/j.cpc.2006.03.003.

[47] K.M. Hangos, A. Magyar, G. Szederkényi, Entropy-inspired Lyapunov functions and linear first integrals for positive polynomial systems, Math. Model. Nat. Phenom. 10 (2015) 105–123, https://doi.org/10.1051/mmnp/201510309.

[48] K. Hangos, J. Bokor, G. Szederkényi, Analysis and Control of Nonlinear Process Systems, Springer, New York, NY, 2004.

[49] K.M. Hangos, A.A. Alonso, J.D. Perkins, B.E. Ydstie, Thermodynamic approach to the structural stability of process plants, AIChE J. 45 (4) (1999) 802–816.

[50] A.N. Gorban, I.V. Karlin, Method of invariant manifold for chemical kinetics, Chem. Eng. Sci. 58 (2003) 4751–4768.

[51] B. Ács, G. Szederkényi, Z.A. Tuza, Z. Tuza, Computing linearly conjugate weakly reversible kinetic structures using optimization and graph theory, MATCH Commun. Math. Comput. Chem. 74 (2015) 481–504.

[52] G. Szederkényi, J.R. Banga, A.A. Alonso, Inference of complex biological networks: distinguishability issues and optimization-based solutions, BMC Syst. Biol. 5 (2011) 177, https://doi.org/10.1186/1752-0509-5-177.

[53] M.D. Johnston, D. Siegel, G. Szederkényi, Dynamical equivalence and linear conjugacy of chemical reaction networks: new results and methods, MATCH Commun. Math. Comput. Chem. 68 (2012) 443–468.

[54] G. Lipták, G. Szederkényi, K.M. Hangos, Computing zero deficiency realizations of kinetic systems, Syst. Control Lett. 81 (2015) 24–30, https://doi.org/10.1016/j.sysconle.2015.05.001.

[55] B. Ács, G. Szederkényi, Z. Tuza, Z.A. Tuza, Computing all possible graph structures describing linearly conjugate realizations of kinetic systems, Comput. Phys. Commun. 204 (2016) 11–20, https://doi.org/10.1016/j.cpc.2016.02.020.

[56] J. Rudan, G. Szederkényi, K.M. Hangos, Efficient computation of alternative structures for large kinetic systems using linear programming, MATCH Commun. Math. Comput. Chem. 71 (2014) 71–92.

[57] M.D. Johnston, D. Siegel, G. Szederkényi, Computing weakly reversible linearly conjugate chemical reaction networks with minimal deficiency, Math. Biosci. 241 (2013) 88–98, https://doi.org/10.1016/j.mbs.2012.09.008.

[58] B. Ács, G. Szlobodnyik, G. Szederkényi, A computational approach to the structural analysis of uncertain kinetic systems, arXiv:1704.08633v1 [math.DS], 27 Apr 2017.

[59] G. Lipták, A. Magyar, K.M. Hangos, LQ control of Lotka-Volterra systems based on their locally linearized dynamics, in: IFAC-PapersOnLine, vol. 49, Elsevier BV, 2016, pp. 241–245, https://doi.org/10.1016/j.ifacol.2016.07.536.

[60] K. Holmström, A.O. Göran, M.M. Edvall, User's guide for Tomlab/Penopt, Tomlab Optimization, 2006.

[61] Y.Y. Cao, J. Lam, Y.X. Sun, Static output feedback stabilization: an ILMI approach, Automatica 12 (1998) 1641–1645.

[62] K.M. Hangos, I.T. Cameron, Process Modelling and Model Analysis, Academic Press, London, 2001, 543 pp.

[63] I.M. Gleria, A. Figueiredo, T.M.R. Filho, A numerical method for the stability analysis of quasi-polynomial vector fields, Nonl. Anal. 52 (2003) 329–342.

[64] Z.A. Tuza, B. Ács, G. Szederkényi, F. Allgöwer, Efficient computation of all distinct realization structures of kinetic systems, IFAC PapersOnLine 49 (2016) 194–200.

[65] C. Conradi, D. Flockerzi, J. Raisch, J. Stelling, Subnetwork analysis reveals dynamic features of complex (bio)chemical networks, Proc. Natl Acad. Sci. USA 104 (49) (2007) 19175–19180, https://doi.org/10.1073/pnas.0705731104.

[66] S. Müller, J. Hofbauer, L. Endler, C. Flamm, S. Widder, P. Schuster, A generalized model of the repressilator, J. Math. Biol. 53 (6) (2006) 905–937, ISSN 1432-1416, https://doi.org/10.1007/s00285-006-0035-9.

[67] A. Császár, L. Jicsinszky, T. Turányi, Generation of model reactions leading to limit cycle behaviour, React. Kinet. Catal. Lett. 18 (1981) 65–71.

[68] H.F. Lodish, Molecular Cell Biology, W. H. Freeman, New York, NY, 2000.

[69] T.M. Yi, H. Kitano, S.I. Simon, A quantitative characterization of the yeast heterotrimeric G protein cycle, Proc. Natl Acad. Sci. USA 100 (2003) 10764–10769.

[70] A. Isidori, Nonlinear Control Systems, Springer-Verlag, New York, NY, 1995.

[71] A. Magyar, G. Szederkényi, K.M. Hangos, Quasi-polynomial system representation for the analysis and control of nonlinear systems, in: Proceedings of 16th IFAC World Congress, Prague, Czech Republic, 2005 (on CD).

[72] G. Szederkényi, N.R. Kristensen, S.B. Joergensen, Nonlinear analysis and control of a continuous fermentation process, Comput. Chem. Eng. 26 (2002) 659–670.

[73] A.I. Vol'pert, Differential equations on graphs, Math. USSR Sbornik 17 (1972) 571–582.

[74] E. Kaszkurewicz, A. Bhaya, Matrix Diagonal Stability in Systems and Computation, Birkhauser, Boston, MA, 2000.

[75] L. Farina, S. Rinaldi, Positive Linear Systems: Theory and Applications, Wiley, London, 2000.

[76] K.S. Narendra, R. Shorten, Hurwitz stability of Metzler matrices, IEEE Trans. Autom. Control 55 (6) (2010) 1484–1487.

[77] R.A. Horn, C.R. Johnson, Topics in Matrix Analysis, Cambridge University Press, Cambridge, 1991.

[78] G. Szederkényi, Computational Analysis of Nonnegative Polynomial Systems, Scholar's Press, Saarbrucken, 2014 (DSc dissertation, Hungarian Academy of Sciences).

[79] G.B. Dantzig, M.N. Thapa, Linear Programming 1: Introduction, Springer-Verlag, New York, NY, 1997.

[80] G.B. Dantzig, M.N. Thapa, Linear Programming 2: Theory and Extensions, Springer-Verlag, New York, NY, 2003.

[81] G. Belov, On linear programming approach for the calculation of chemical equilibrium in complex thermodynamic systems, J. Math. Chem. 47 (2010) 446–456.

[82] S. Kauchali, W.C. Rooney, L.T. Biegler, D. Glasser, D. Hildebrandt, Linear programming formulations for attainable region analysis, Chem. Eng. Sci. 57 (2002) 2015–2028.

[83] V. Gopal, L.T. Biegler, Nonsmooth dynamic simulation with linear programming based methods, Comput. Chem. Eng. 21 (1997) 675–689.

[84] J.A. Klein, D.T. Wu, R. Gani, Computer aided mixture design with specified property constraints, Comput. Chem. Eng. 16 (1992) S229–S236.

[85] G.L. Nemhauser, L.A. Wolsey, Integer and Combinatorial Optimization, John Wiley & Sons, London, 1999.

[86] J. Löfberg, YALMIP: a toolbox for modeling and optimization in MATLAB, in: 2004 IEEE International Symposium on Computer Aided Control Systems Design, IEEE, 2004, pp. 284–289, Available from: http://www.researchgate.net/profile/Johan_Loefberg/publication/4124388_YALMIP__a_toolbox_for_modeling_and_optimization_in_MATLAB/links/00b49519c965559d32000000.pdf.

[87] A. Makhorin, GLPK 4.9, 2006, Available from: http://www.gnu.org/software/glpk/glpk.html.

[88] K. Homlström, M.M. Edvall, A.O. Göran, TOMLAB for large-scale robust optimization, in: Nordic MATLAB Conference, 2003.

[89] A. Bemporad, M. Morari, Control of systems integrating logic, dynamics, and constraints, Automatica 35 (1999) 407–427.

[90] H.P. Williams, Model Building in Mathematical Programming, Wiley, London, 1993.

[91] S. Boyd, L. El-Ghaoui, E. Feron, V. Balakrishnan, Linear Matrix Inequalities in Systems and Control Theory, SIAM Books, Philadelphia, PA, 1994.

[92] H.D. Tuan, P. Apkarian, Y. Nakashima, A new Lagrangian dual global optimization algorithm for solving bilinear matrix inequalities, Int. J. Robust Nonl. Control 10 (2000) 561–578.

INDEX

Note: Page numbers followed by *f* indicate figures, *t* indicate tables, and *b* indicates boxes.

A

Affine transformation, 38–39
Autonomous dynamical model, 9–11
Autonomous ordinary differential equation, 37

B

Bilinear matrix inequality (BMI), 97–98, 130*f*, 147
   stabilizing control of quasipolynomial system, 95–99
   time-reparametrization problem as, 58–62
Bio-chemical reaction network (Bio-CRN), 5–6, 29, 114
   computation of linearly conjugate, 85–86
   dynamical equations of, 29–31
   reaction graph, 31*f*, 33*f*
Bio-reactor model, 5–6
BMI. *See* Bilinear matrix inequality (BMI)

C

Canonical mechanism, 25
Canonical realization, of ODE system, 26*f*
Chemical kinetic scheme, 44
Chemical reaction network (CRN), 1–2, 5, 24, 38, 111–112
   bio-CRN, 29
     computation of linearly conjugate, 85–86
     dynamical equations of, 29–31
     reaction graph, 31*f*, 33*f*
   embedding, 52–53
   Kirchhoff matrix of, 19–20
   model, 16, 19
     Lotka-Volterra (LV) model, 36*f*
   ordinary differential equation structure of, 29–30
   positive diagonal transformation of, 39–43
   with rational functions as reaction rates, 5–6
   realizations of closed-loop system, 99–104
   structure, 16–17

Closed-loop system
   complex balanced realization of, 106*f*
   feedback computation in complex balanced, 101–102
Column conservation matrix, 19–20
Compartmental matrices, 142–143
Computational analysis of structure, of kinetic system, 70–71
Computational tools, linear programming, 144–145
Convex polytope, 153
Core reactions, 80
CRN. *See* Chemical reaction network (CRN)

D

Deficiency one theorem, 63
Deficiency zero theorem, 23, 62–63, 79
Diagonal stability, LQ with, 94–95
Diagonal transformation
   nonlinear, 44–47
   positive, 39
Directed graphs, 139–141
Dynamical equation, of bio-CRN, 29–31
Dynamical models
   application of, 1
   autonomous, 9–11
   behavior, 1–2
   notion and significance of, 2–3
   uncertainty in, 152–153
Dynamical system theory, 3

E

Embedding transformations, 50
   embedding rational functions into polynomial form, 52–53
   embedding smooth nonlinear models into QP form, 50–52
Equivalence classes of quasipolynomial system, 47–50

F

Feedback control
  LQ-based state, 92
  of nonnegative polynomial system, 99–104
Feinberg-Horn-Jackson graph, 18
Fermentation process, in quasipolynomial form, 126–128
Fermenter model, variables and parameters of, 127*t*, 129*t*, 133*t*
5-node-repressilator analysis, with auto activation, 111–113, 111*f*, 113*f*
Full reaction kinetic model, 21

G

Generalized Lotka-Volterra system, 1–2, 7–8
G-protein (guanine nucleotide-binding protein) cycle, 119, 120*f*, 121–122, 140–141
  heterotrimeric, 119
  model, 120
G_1/S transition model, in budding yeast, 108–109, 110*f*

H

Heterotrimeric G-protein, 119
Hurwitz stable matrix, 142

I

Input-affine nonlinear state-space models, 148–149
  stability analysis, 149–150
  stabilizing feedback controllers, 150
  transformation of states, 148–149
Input-affine system model, 151
Input-output linearization via state feedback, 151–152
Invariants, for quasipolynomial and kinetic systems, 65–68
Irreversible Michaelis-Menten mechanism, 21, 21*f*, 26–27, 26*f*, 31*f*
Iterative BMI algorithm, 131–132
Iterative LMI (ILMI) algorithm, 97–98, 129–130, 130*f*

J

Jacobian matrix, 61
  in LV case, 57
  of quasipolynomial system, 56

K

Kinetic models. *See also* Uncertain kinetic models
  computation-oriented representation, 86–90
  dense realizations, 89–90
  full reaction, 21
  with mass action reaction rates, 16–34
    as QP models, 34
  physical-chemical origin of natural Lyapunov function of, 68–69
  and QP model, relations between Lyapunov function, 68–70
  with rational reaction rates, 16–34
  sparse realizations, 89–90
  uncertainty in, 87
    realizations of, 87–89, 90*b*
Kinetic networks, with mass action law, 3–6
Kinetic system, 1–2
  computational analysis of structure of, 70–71
  conjugate realizations of, 71–72, 75–80
    complex balanced realizations, 77
    deficiency zero realizations, 77–79
    optimization framework, 79–80
    reaction graph, 79*f*, 82*b*, 84*f*
    weakly reversible structures, 76–77
  dense linearly conjugate realization of, 74*b*
  dense reaction graphs, 72–74
  dense realizations, 72, 85*f*
    CRN structures using, 75
  dynamic description of, 3–4
  graph structures of, 80–85
  invariants (first integrals) for, 65–68
  Lotka-Volterra (LV) model as, 34–36
  notations for, 137–138
  plausible, 4–5
  sparse realizations, 74–75
    CRN structures using, 75
  stability of, 62–65
Kirchhoff matrix, 19–20, 35–36, 74–76, 85–86, 102–103, 115–120, 143

L

Linear conjugacy, 39–43, 71, 78–79
  of CRN with rational reaction rates, 41–43
  of networks with mass action kinetics, 40–41
Linear first integrals, 67
Linearized-quasipolynomial model, 93
Linear kinetic system, 67
Linear matrix inequality (LMI), 57–58, 61–62, 94–95, 147
  global stability analysis, 57–58
Linear programming (LP), 144–145
  computational tools, 144–145

Printed in the United States
By Bookmasters